Patterns of Sexuality and Reproduction

ALAN S. PARKES

Patterns of
Sexuality and Reproduction

OXFORD UNIVERSITY PRESS
LONDON   OXFORD   NEW YORK
1976

Oxford University Press, Ely House, London W.1

GLASGOW NEW YORK TORONTO MELBOURNE WELLINGTON
CAPE TOWN IBADAN NAIROBI DAR ES SALAAM LUSAKA ADDIS ABABA
DELHI BOMBAY CALCUTTA MADRAS KARACHI LAHORE
KUALA LUMPUR SINGAPORE HONG KONG TOKYO

© OXFORD UNIVERSITY PRESS 1976

First published in the Oxford Paperbacks University Series 1976 and simultaneously in a cloth bound edition.

Parkes, Alan Sterling
 Patterns of Sexuality and Reproduction

 Bibl. – Index
 ISBN 0-19-217652-8
 ISBN 0-19-289081-6 Pbk.
 1. Ti
 DC 612.6
 LC HQ 31
 LCSH Sex

Set in Great Britain by Gloucester Typesetting Co Ltd
Printed by Fletchers Limited, Norwich

Preface

In 1966 I published a collection of lectures, addresses, and articles under the title *Sex, Science and Society*, a title which seemed to cover the mixed bag which made up the volume. The bag was indeed mixed—an autobiographical sketch of my early days, and lectures on the scientist and his work, the development of various personal researches, population problems, and fertility control, the functions of taste and smell in animals, low-temperature biology, and a number of miscellaneous topics. At the time, it seemed a good idea to leaven the loaf in some way and I persuaded Alfred Wurmser, who I had met previously, to illustrate the volume with cartoons designed to indicate, with a few strokes of a pen, ideas which might otherwise involve a lot of words. Wurmser did this most successfully, and many of his drawings were an unusually attractive form of visual aid. The others were evidently meant to deflate the scientist, and did so in a kindly fashion.

This innovation was well received in some quarters, but unfortunately, on first handling, the book gave an impression of levity which on the whole seemed to evoke the stern rebuke that you should not do that to the scientist. Later on, an old friend, Alan Morton, formerly Professor of Biochemistry in the University of Liverpool, said to me 'You ought to know that, contrary to the laws of nature, scientists rise by gravity and fall by levity.'

Things have changed since then and the permissive age seems to have penetrated even the ivory towers of science, as witness the suggestion in *Biologist* (the Journal of the Institute of Biology) that operations to restore the physical signs of virginity might infringe the Trade Descriptions Act. Even so, on the principle of never re-lighting a cigar, it seemed desirable to use some other form of visual aid in the present volume, especially as the subject matter lent itself ideally to diagrammatic illustration.

In the Foreword to *Sex, Science and Society* I asked whether the addition of a volume of collected lectures, already mainly published, to the torrent of current publications could be justified. My justification then was a purely personal one—I had a lot of fun doing it. My justification now is more valid —I do not know of any existing volume dealing with human reproduction in this particular way, and designed to be intelligible to the educated layman, of interest to the non-specialist biologist, and perhaps even to provide a few crumbs for the expert on reproductive biology. At least, the subject matter may supply some background information to those who think about human reproduction, its processes, and social consequences, as well as merely practise it.

In such a vast field it is difficult to avoid superficiality, and it has seemed better to deal in some detail with typical examples of our knowledge rather than to give what would have amounted to little more than a catalogue of the literature. Even so, it has been possible to include the bibliographic references to only the more important publications and to the sources of the illustrations. Many statements have had to be made without citing the authority for them, and I must apologize to the many authors whose work I have used but whose names and references have not been included. Moreover, by nature of the volume, the references cannot be up-to-the-minute, but in the context this is not important. The biometry of human reproduction is not a field in which epoch-making discoveries are being made, and for the most part the material used deals with well-established patterns. The volume, therefore, even if not up-to-date, is not out-of-date.

There have been the usual problems inseparable from assembling an anthology of this kind—levelling up treatment, eliminating overlap, and so on. With so much to draw on, the selection of material has necessarily been highly esoteric, especially with the references. Thus the temptation to go on revising by adding and deleting was almost irresistible, but if the work was to be of use or interest to anyone else, it had to be finalized, and here is the result.

London   A.S.P.
January, 1976

Contents

Acknowledgements

My first acknowledgement must be to Mrs. Doreen E. King, Dee to her friends, who for sixteen years has given me invaluable assistance. Mrs. King was responsible for much of the library research involved in the preparation of these essays, and in one way or another has been involved in all the work which has gone into them. My warmest thanks to her.

I am also much indebted to Mrs. Margaret A. Herbertson, Assistant Editor of the *Journal of Biosocial Science*, for friendly help in assembling this volume and in proof-reading.

It is a pleasure to express my indebtedness to my many other friends and colleagues who have helped in one way or another in the preparation of the essays. Chapter 8, for instance, owes much to discussions with Dr. W. H. James. I am grateful also to the staff of the Oxford University Press, who gave me the final impetus to complete the work.

The lectures and articles upon which these essays are based may be regarded as the first fruits of a grant made to me in 1966 by the World Health Organization to assist the preparation of a monograph on *Individual and ethnic variation in human reproductive function*, a task still incomplete. My thanks to that Organization.

I am indebted to the editors or publishers of the journals and books listed in the References for permission to make use of the material, to many authors for permission to reprint diagrams, and to the Librarian of the Royal Society of Medicine and his staff for much help.

I

The reproductive life-cycle

Reproductive function is characterized in all mammals by a series of cycles—of which the major one is the life-cycle of immaturity, puberty, maturity, menopause, and finally senility and death. The proportion of the life-cycle occupied by the reproductive phase varies from species to species and from individual to individual of the same species; it is difficult to generalize, for not all humans—and probably few animals—survive to die of old age, or even to reach the end of their reproductive life.

In man, there is a marked sex difference in the span of reproductive life. The end-points are much more definite in the female than in the male: the beginning and end of the female's menstrual cycle give a fairly clear indication of the maximum extent of her reproductive life. There are no such clear indications in the male; the onset of both puberty and of the climacteric—when sexual activity and potency decline —is less overt and more gradual. However, the male usually retains fertility long after it has disappeared in females of similar age. One may well doubt reports of centenarians siring offspring, but there are well-authenticated records of men of 70–80 years of age becoming fathers. By contrast, though women as a whole live longer than men, expectant mothers of 50 years of age are rare.

The reproductive organs

In mammals, the primary organs of reproduction, the gonads (testes in the male and ovaries in the female), are essentially dual-purpose organs of both external and internal secretion.

The testes produce spermatozoa and the androgenic hormones (mainly testosterone), which induce and maintain the attributes of maleness. The ovaries produce ova and the oestrogenic and progestational hormones (mainly oestradiol and progesterone), which are concerned with the development and maintenance of the attributes of femaleness, including sexual receptivity and the inception of pregnancy.

Associated with the gonads are accessory organs which play a part, mostly an essential part, in the reproductive processes. In the male, these consist of the scrotum, which houses the testes, the two epididymides, which collect the spermatozoa from the testes and in which the spermatozoa mature, the two ducts, which convey them to the urethra and thence to the exterior, the seminal vesicles and prostate glands, the secretions of which dilute the sperm mass, and the penis through which the urethra passes and which conveys the semen to the female.

In the female, the accessory organs consist essentially of the female reproductive tract leading from the ovaries to the exterior—the Fallopian tubes, uterus, cervix, vagina, and vulva, with its associated labia and clitoris. Usually essential also are the mammary glands (from which the mammals take their name), though the human breast in sophisticated societies has largely been rendered superfluous by artificial feeding, and seems likely to degenerate into a mere psychosexual symbol. During the reproductive life-cycle, the gonads and accessory organs show various changes.

Secondary sexual characters

Many animals—in addition to the essential reproductive organs, gonads, and accessory organs—possess secondary sexual characters which, although mainly gonad-dependent and showing highly significant changes during the life-cycle, play no part (except perhaps as visual stimuli) in the reproductive processes. Where present, such characters are usually better developed in the male than in the female; in such cases, it is then possible, as in man, to distinguish the two sexes without reference to the external genitalia—a fortunate state of affairs in civilized society.

In man, for instance, the mature male can usually be distinguished from the female by the presence of facial hair, a different pattern of pubic hair, a deeper voice, and a larger and more muscular body. All these are dependent on the androgenic substances produced by the testes, in the absence of which sexual differentiation of this kind does not appear. The female pattern of pubic hair, though not gonad-dependent, being found in youths, castrated men, and in ovariectomized women, is nevertheless dependent, as is the axillary hair, on maturation of the body at puberty. This neutral pattern of pubic and axillary hair appears to be dependent on androgenic substances from the adrenal glands.

Six stages of pubic hair growth can be distinguished, of which the first five lead to the formation of the inverted triangle of coarse curly hair common to adolescents of both sexes. Stage 6 appears some years after stage 5 and is characterized by the upward extension of the hair to a point at the navel. This stage is found in 80 per cent of mature men but in only 10 per cent of mature women.

Voice-changes in the boy at puberty result in the pitch deepening from treble or alto to tenor or bass. The immediate cause is the enlargement of the larynx and lengthening of the vocal chords which accompanies the spurt in somatic and penis growth. The voice-changes at puberty, being androgen-dependent, tend to regress in old age.

The amount and distribution of head hair is of course greatly influenced by genetic factors; but baldness, which appears so often in early middle age, is not seen in eunuchs and evidently results from some action, probably on the scalp, of the male hormones. Skin changes at puberty are well known, and congestion of the sebaceous glands caused by androgens, leading to a greater or lesser degree of acne, is frequently seen. At the other end of the life-cycle thinning and wrinkling of the skin due to loss of elasticity and atrophy of subcutaneous tissue are well-known effects of ageing.

As to body configuration, the smaller size of the female and the subcutaneous fat which causes her rounded contours are due at least in part to the bone-ossifying and fat-mobilizing

effects of oestrogen and, together with the wider pelvis, constitute the chief positive secondary sexual characters of
women—those that are caused by femaleness rather than by
the absence of maleness. In contrast, the greater musculature
of the male is caused by the tissue-building action of certain
of the androgens.

Maturation of the male

Five stages are recognized in the maturation of the external
genitalia of boys: slow development from the infantile condition; enlargement of scrotum; enlargement of penis; 'sculp-
turing' and darkening of penis; and lastly the appearance of
the adult condition. In a sample of American boys, it was
observed that stage 2 started on average at about 12 years of
age and that the time-lag between stages 4 and 5 was much
greater than between other stages—stage 5 being reached at
about 17½ years of age. The epididymis and accessory glands
show appreciable activity at birth as a result of being subjected to maternal hormones during the later stages of foetal
life. Post-natal regression then leads to a state of quiescence
which lasts until puberty, when development starts at about
10 years of age and keeps pace with that of the testes.

The testes are normally fully descended early in life, but
they increase little in size during the first decade. They then
undergo relatively rapid growth which takes them to full size
at the age of about 20 years. Further slight increases in size
may occur during adulthood, and there is little decrease even
in extreme old age. These size changes are conditioned largely
by the seminiferous tubules which produce the spermatozoa.
The intertubular cells develop to a similar time-table but
contribute little to the overall bulk of the testes.

The interstitial tissue is probably responsible for the production of most, but not all, of the androgen which is excreted
in the urine—the remainder coming from the adrenals, two
small glands above the kidneys. Total excreted androgen
shows a very clear rise and fall during reproductive life
(Fig. 1.1). This and similar curves have given rise to the
notion that, androgenically, a man passes his prime at about

30 years of age; functionally, however, androgen is adequate for much longer, as shown by the fact that the rise in the urinary secretion of gonad-stimulating substance, indicating loss of androgen control of the pituitary gland, starts only around 60 years of age.

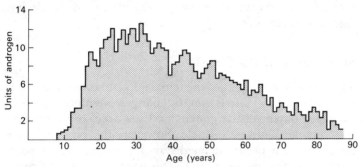

FIG. 1.1.   Excretion of biologically active androgen by boys and men.[38] By this criterion, men reach their endocrinological peak of masculinity about age 30 years and then go continuously down hill.

It is not easy to determine the time when spermatozoa are first produced in any particular individual, especially as it may precede the capacity for ejaculation. However, the cells lining the tubules which produce the spermatozoa begin to show activity at about 10 years of age and active spermatogenesis starts some 3 years later. Once started, spermatogenesis normally proceeds vigorously throughout adulthood and often into old age. This continuous production of vast numbers of germ cells by the male is in sharp contrast to the state of affairs in the female.

The physical signs of sexual capacity in the male are erection, orgasm, and ejaculation, but the three are not necessarily connected. Erection and orgasm can take place from an early age—even in infants. The capacity to ejaculate appears much later and may first be shown by nocturnal emissions, often without erection. From the data given by Kinsey and his colleagues[24] it seems that 50 per cent of his patients had their first ejaculation at about 13 years and 10 months. Much earlier and much later first ejaculations

(8–21 years) have been recorded, without any suggestion of physiological abnormality. At the other end of the life-cycle, the capacity for erection and ejaculation is usually retained into old age; Kinsey's records show that about 75 per cent of American men were potent at 70 years of age, but only 20 per cent of octogenarians. However, ejaculation is much less vigorous in old men.

Maturation of the female

The sexual maturation of the female, like that of the male, is characterized by various somatic changes, but more especially by development of the gonads and the accessory reproductive organs. Unlike the male, however, the process in the female has one clearly defined stage: menarche, the appearance of the first menstruation. This has served as a reference point for almost all investigations into female puberty and, by the comparative ease with which it can be noted and recorded, has tended to obscure the importance of other events such as the first ovulation, development of the breasts, and changes in body configuration.

Mammary development (development of the breasts) is an obvious feature of puberty in girls, but it is difficult to quantify. Five stages of development have been distinguished. The first stage involves elevation of the nipple. Stages 2 and 3 are signalled by an increasing elevation of breast and nipple to form a small mound, together with extension of the surrounding pigmented skin—the areola. In stage 4, the areola and nipple project to form a secondary mound above the level of the breast. By stage 5 only the nipple projects and maturity has arrived. Of these five stages, the first usually appears before menarche and the last between 14 and 15 years of age. During the pubertal changes in the vagina and external genitalia, the vagina enlarges and its secretions become more acid. The clitoris becomes erectile and the labia enlarge.

Following slow enlargement during early childhood, the uterus starts on more rapid development around the age of 10 years and increases in size by many times in the next few years under the influence of ovarian oestrogens. Finally, with or without a preliminary ovulation, a sudden change in hormonal and neural stimuli causes breakdown of the lining of the uterus with resulting effusion of blood. This drains to the exterior through the cervix and vagina, producing the first menstrual bleeding.

The age at which the first menstruation occurs varies considerably from individual to individual, but less so in different populations than is commonly supposed (see p. 17). There is another factor. It is well known that the age of puberty in Europe and the United States has decreased substantially in modern times. A figure of 1 year every 10 years is sometimes quoted, but this alarming estimate is much exaggerated. More likely, the age of menarche has decreased by about 4 years in the last century. Fig. 2.2 (p. 18) implies a reduction of only about 3 years in the last century. This lowering in the age of onset of menstruation is presumably due to the improvement in environmental conditions—it is reminiscent of the observation that better physical development in boys is associated with earlier puberty.

Contrary to popular belief, the length of the menstrual cycle, even when well established, is very variable (see p. 24). The variation does not show any very regular pattern during the major part of a woman's reproductive life, though it is often irregular in the years following menarche and during the menopause. Extensive studies of variation during the life-cycle have revealed a decline in length in the 6 post-menarche years with a high variability, a slight decrease between the ages of 20 and 40 years with reduced variability, and a sharp increase in both length and variability in the 8 pre-menopausal years. This irregularity is contributed to by the occurrence of cycles in which ovulation fails and by incomplete cycles. According to Döring (Fig. 1.2) both have a much increased incidence at the beginning and towards the end of reproductive life.[13]

The ovaries undergo far less absolute increase in size during sexual development than do the testes. According to published figures, they enlarge to a maximum size between 20 and 40 years of age and thereafter decrease to about half-size.

The most interesting fact about the ovaries, however, is that their supply of eggs is laid down very early in life and is not added to thereafter. The size of the original store is not easy to assess, but it may be around half a million in the human female. This number is greatly reduced by decay of follicles (egg-storing sacs) even before puberty and the process continues throughout reproductive life, so that by the time of the

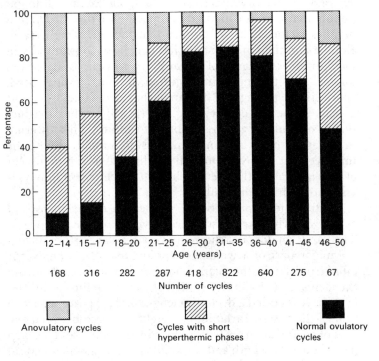

FIG. 1.2.   Anovular, incomplete, and normal menstrual cycles in girls and women.[13] Both categories of atypical cycle are more frequent in girls and older women, normal cycles being most frequent at ages 25–35 years.

menopause the supply of ovarian eggs may be entirely exhausted. By contrast, the loss of ovarian eggs by ovulation is negligible: even assuming the regular discharge of one egg each 28 days for 30 years, the total is only around 400 eggs.

The laying down of the ovarian ova very early in life has important practical implications: it means that during reproductive life the ova are anything between 15 and 45 years old—and that a woman aged 45 years who produces a child does so from an ovum of that age. This fact may perhaps account for the greater number of abnormal foetuses produced by older mothers (Fig. 1.3)—though there is the alternative explanation that the decreased frequency of intercourse among middle-aged couples may result in the ovum or spermatozoon being 'stale' at the time of fertilization and thus more likely to produce an abnormal foetus. This second explanation would account for the slight effect of paternal age on the abnormality rate. Probably both factors are involved.

The release of a single ovum during each cycle is, of course, the rule for women; and, by analogy with other animals that give birth to single offspring, it is likely that the two ovaries ovulate alternately. However, a single egg may result in twins by splitting of the embryo at an early stage of development. Such 'monozygotic' twins, commonly called identical twins because of their typically identical genetic make-up, form only a small proportion of all twins. The majority of twins result from the simultaneous, or near-simultaneous, ovulation of two eggs to produce 'dizygotic' twins. From the point of view of the life-cycle, the interesting thing is that the incidence of monozygotic twins shows no significant connection with the age of the mother, whereas the frequency of dizygotic twins shows a very marked correlation—rising by a factor of 5 or so to a peak in mothers between 30 and 40 years of age (Fig. 1.4). It is arguable therefore, that the ovaries reach their maximum biological activity, as well as their maximum size, comparatively late in reproductive life.

The menopause

The menopause is very variable in its time of onset and has no definite end-point—the cycle appearing to fade out rather than stop—and so the process is difficult to time accurately, even in a single individual. Many factors have been said to influence the age at menopause, including ethnic group, age at menarche, socio-economic circumstances, occupation, urban or rural environment, and marital and maternal history. It is difficult to know whether there is currently a trend towards later menopause as there is towards earlier menarche; certainly on the evidence available it seems unlikely (see p. 20).

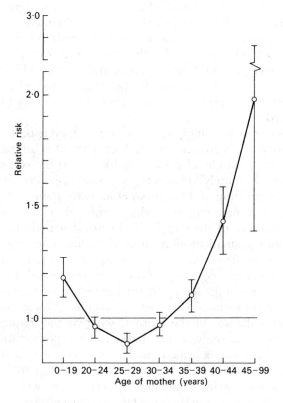

FIG. 1.3.   Congenital abnormality according to age of mother.[35] The risk is least at ages 25–29, and increases rapidly up to the end of reproductive life. Vertical lines indicate the range of possible values.

Functionally, the menopause is characterized by increasing irregularity of menstruation, increasing failure of ovulation, and increasing secretion of pituitary gonadotrophin (in response to the massive reduction in oestrogen secretion by the ovary). To what extent the onset of the menopause is determined by exhaustion of the eggs in the ovary, by failure of hormone production in the ovary, or by failure of the overall mechanism controlling reproductive function is uncertain.

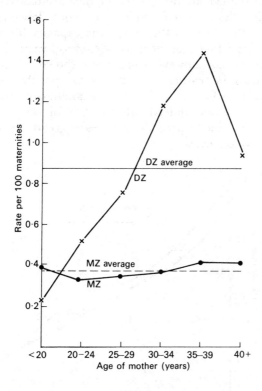

FIG. 1.4. DZ (two-egg) and MZ (one-egg—identical) twinning rates according to age of mother.[28] The DZ rate rises sharply with increasing age, the MZ rate is comparatively constant at a much lower level.

Sexuality

So far I have considered the development and regression of sexual organs and of sexual function during the life-cycle in the two sexes separately. Fertility, however, in the sense of the production of viable young, involves the operation of coitus— the characteristics, frequency, and effectiveness of which change markedly during the life-cycle. In the male, reaction time increases with increasing age, and the subjective characteristics of the ejaculatory process change gradually. In the female—in whom sexual response is more dependent on experience than it is in the male—the capacity for orgasm may increase rather than decrease for several years. The frequency of coitus, however, has been shown to decrease steadily with increasing age (Fig. 1.5)—possibly waning interest in sex is further attenuated by familiarity. Effectiveness also decreases; indeed, the number of acts of coitus required to produce a pregnancy increases greatly during the reproductive life cycle.

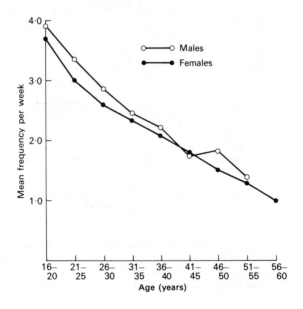

FIG. 1.5.   Frequency of marital intercourse according to age of husband and wife.[24] The higher level for men presumably derives from the husbands being slightly older than their wives, thus displacing the husbands' curve to the right.

At best, the probability that conception will result from a single act of coitus is small, and the advent of effective contraception has further dissociated sexuality and reproduction. Let us hope that this dissociation will not be completed by the development of ectogenesis—production of a child outside the mother.

2

Variation in reproductive function

Individual variation is inherent among living organisms and this generalization is nowhere truer than in human beings. In man, the situation is further complicated by the addition of social vagaries to biological variation; in reproductive function, especially, potential must be distinguished carefully from performance. Thus most girls become fertile between the ages of 12 and 14 years, but whether this potential is exercised then or later depends on the social *mores* ruling at the time and place in question. Lactation is another function which shows great biological variability, but is further diversified by social customs, while seasonal variation in the birth-rate in man is determined largely by social rather than biological factors. In contrast, the length of gestation shows the usual biological variation, but is little influenced directly by social factors.

Individuals, then, vary even within a comparatively homogeneous racial community, but they may also be of very different ethnic groups and possibly show ethnic, in addition to individual, variation in reproductive potential and performance. Further, they may live in the tropics or within the Arctic circle, and conditions today may be very different from what they were 100 years ago. It is not easy, therefore, to disentangle the various factors involved; the difficulty is particularly great in the case of possible ethnic variation, the conclusive demonstration of which would require the comparison of different ethnic groups living side by side in large numbers under similar conditions. In the past such situations have been rare, but with the present drift towards multiracial societies, they will doubtless become more frequent until the picture is blurred by increased miscegenation. In the meantime, one can rely only on a balance of probability. For instance, while conclusive proof is not available, it is difficult to avoid thinking that the great variation in the dizygotic twinning rate, from 0·2 per cent in Japan to more than 4 per cent in parts of Africa, has a genetic component.

Puberty to climacteric in men

The changes that constitute puberty in boys, outlined in Chapter 1, are too gradual for accurate measurement, and the age of puberty depends on what stage of what characteristic is considered diagnostic. Whichever criterion is taken, however, there is much individual variation (Fig. 2.1·).

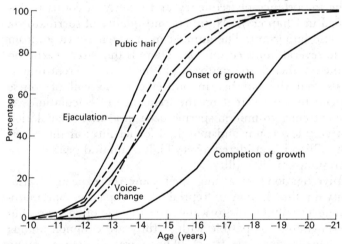

FIG. 2.1. Age of appearance of various indices of puberty in American boys.[24] These data imply that 50 per cent of boys develop pubic hair at 13 years 6 months, ejaculate at 13 years 10 months, experience the voice-change at 14 years 4 months, and complete growth at 15 years 5 months.

Ethnic variation is less certain. There is a credible report that the median age of appearance of pubic hair, voice-changes, and first ejaculation is not significantly different in white and negro boys in the United States. By contrast there is said to be a difference of nearly 2 years between the time of the first appearance of pubic hair in Israeli boys of Middle East and of European extraction, and there are other similar records. The onset of fertility is also variable. Sperm-counts may be low in the pubescent male, though it has been reported that 14–15 year old boys may produce spermatozoa normal in numbers, quality, motility, and morphology, so that they should be fertile. Some certainly are. Fathers of 14 years of age, implying fertile intercourse at 13–14 years, have always been on record; the present increase in their numbers probably owes more to social than to biological factors.

The reproductive competence of adult males is equally diverse. Sexual activity may take many forms, but only coitus with ejaculation can be regarded as potentially reproductive. (The importance of frequency and quality of coitus is discussed in Chapter 7.) Numbers and quality of spermatozoa are also important. The number in an ejaculate varies from nil to several hundred million. There is thus great variation in the number of spermatozoa produced by different individuals, but the number in one ejaculation will of course depend to some extent on the frequency of ejaculation. At one extreme, 20 million spermatozoa in a single ejaculation is a very low count and marginal for fertility; at the other, 500 million is considered a very high count and gives at least a presumption of fertility.

Investigations on ageing men show the expected variability in the decline of reproductive capacity. Spermatogenesis usually begins to wane after the age of 60 years, but sperm-counts compatible with fertility are common among septuagenarians (see p. 5). There are, of course, many reports of the ability of very old men to sire children. Some of these must be considered equivocal, especially the reports of numerous fertile centenarians in parts of Russia. Some, however, are possibly authentic, as for instance the account of the American farmer, aged 94 years, who is said to have produced sixteen children during his first marriage, and a seventeenth in his 94th year, by a second wife. There is thus great biological variation in the retention of fertility in old men, and its demonstration depends on personal and social factors which defy assessment.

Menarche to menopause

Most of the changes occurring at puberty in the female are gradual and as difficult to assess as those occurring in the male; even the one sharply defined change, the appearance of the first menstruation, may be misleading in that it does not necessarily presage the start of regular cycles with ovulation. Moreover, in quantitative studies various problems arise, the first of which is how to assess the age of menarche. Ideally, a group of girls should be followed closely from a pre-pubertal age so that the appearance of the first menstruation can be accurately recorded to within a few days. Long-term studies of this kind, however, are rarely possible, and two closely related methods have mainly been used. In the recollective method the girls are asked at what age they started to menstruate, and the result is expressed as the average age. In the cross-sectional method groups of girls of different ages are surveyed to find out how many have started to menstruate in each age group. The result is assessed in terms of the age group at which some given percentage, usually 50 per cent, are menstruating. Curves constructed on this basis are typically S-shaped, as is usual with dose–response curves, dose in this case being the age and response the percentage menstruating.

The curves shown in Fig. 2.2 indicate a variation from 10 years to 16 years in modern times, with the 50 per cent mark at a little under 13 years. For instance, observations on some 1500 Dutch girls showed that menarche occurred between 13 and 15 years of age in about 60 per cent of the girls, with the 13½–14 year old bracket including no less than 50 per cent of these. These observations accord with many results obtained in places as diverse as Brazil, Sante-Fe, India, and the Phillipines. There is plenty of evidence here of individual, but little of ethnic variation.

It is likely that where higher mean ages of menarche have been recorded, as for Eskimos and Bantus (both over 15 years of age) (see Fig. 2.3), they imply not ethnic variation but a social and nutritional environment analogous to that existing in England in the middle of the nineteenth century when the age of menarche was considerably higher than it is now (Fig. 2.2). Many other factors may also be involved, such as social class, urbanization, season of the year, and altitude.

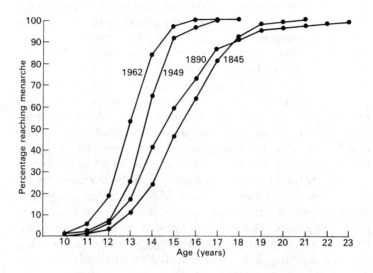

FIG. 2.2.   Changes in the age of menarche in England.[5] In 1962, 50 per cent of girls had started to menstruate by 13 years of age. A century earlier the corresponding figure was a little over 15 years. Also, as shown by the slope of the curve, the variation was greater in the nineteenth century, from 11 years to 19 years compared with 10–16 years for girls in the 1960s.

The onset of fertility in relation to menarche is of obvious importance in this permissive age. Although adolescents may be relatively infertile as a result of anovular or incomplete cycles, it cannot be too strongly emphasized that, if the first cycle is a complete one, the first ovulation will occur before the first menstruation, so that conception is possible at puberty without menstruation having started. This has, in fact, been reported by Döring. The record for juvenile fertility, according to press reports, is currently held by a Swedish girl who was found to be pregnant with triplets at the age of 13 years. The father was said to be 16. Even this impressive performance, like the first 4-minute mile, will no doubt be improved upon.

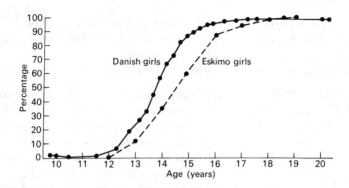

FIG. 2.3. Age at menarche among the Eskimos in south-west Greenland compared with that of Danish girls in Copenhagen.[2]
Eskimo girls are about 2 years older than Danish ones at the
stage when 50 per cent are menstruating.

In this field, however, individual records are less interest-ing than mass performance, and the relation between men-arche and fertility was studied on a large group of girls in India at a time when active contraception in that country must have been minimal. The most common ages of men-arche were 13, 14, and 15 years (ranging from 10 years to 19 years); the most common age at consummation of mar-riage was 16, 17, or 18 years (ranging from 11 years to 26 or more years); and first conception had a peak extending from 16 years to 19 years (ranging overall from 11 years to 16 years or more). As might be expected the first conception followed closely on the consummation of marriage.

So much for variation in female adolescence. Variation is even greater at the end of the reproductive life-cycle—the menopause—and as the menstrual cycle tends to tail off rather than stop abruptly, the variation is made even more difficult to assess by the lack of a definite end-point. As a result, various surveys have produced various results. Many of the questions were, however, dealt with authoritatively in a survey carried out in 1965. The investigators[31] studied the records of 736 women aged 45–54 years living in or near London, using the following definitions:

Pre-menopausal: menstruation within the last 3 months
Menopausal: menstruation within the last year but not in the last 3 months
Post-menopausal: no menstruation within the last year

The average age at the natural menopause was 50·78 years. Contrary to some other reports, no conclusive evidence was obtained of any increase in the age at menopause over the last century. The menopause was found to occur later in currently married women than in single or previously married women. Considering currently married women only, increased parity (a larger number of children) was found to be associated with a later menopause in women of higher but not of lower socio-economic class. Other modern estimates of the age at natural menopause are 50·1 years (Scotland), 51·4 years (Netherlands), and 49·8 years (United States). In the West, therefore, it would seem that the usual age at menopause may be 50 years or a little more, but figures given for London[31] show that a small percentage of women may still be pre-menopausal at the age of 53 years or post-menopausal at the age of 45 years (Fig. 2.4). Comparable figures for Eastern races would be of interest.

The cessation of fertile life, however, does not necessarily correspond with the last menstrual bleeding, because it is likely that there is an increasing number of anovular cycles or other inhibitions of fertility as the menopause approaches. Thus, according to Fig. 4.4 (p. 69), only a few births per 1000 women aged 40–44 years took place in America and Sweden in 1966. Even the super-fertile Hutterites, who eschew contraception, produced only about 50 births per 1000 women in the age group 45–49 during the period 1936–40. According to figures for England and Wales between 1938 and 1968 only about 50 thousand legitimate births out of a total of 22 million were to mothers of 45 years or more in age.[21] Rather surprisingly about the same proportion (approximately 1:400) occurred to mothers of this age among the nearly 1·4 million illegitimate births. The comparatively large number of births some time ago ascribed to Japanese women at ages over 50 years could have been accounted for by grandmothers registering illegitimate grandchildren as their own.

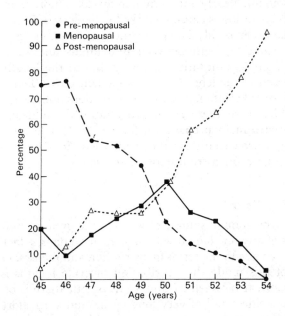

FIG. 2.4.   The age at menopause among women in London.[31] The proportion of menopausal women increases as the proportion of pre-menopausal women decreases up to the age of 50 years, and then decreases as the proportion of post-menopausal women increases.

The menstrual cycle

In women breeding to the limit the menstrual cycle is largely inhibited by successive pregnancies and lactations, and such a situation is still found in primitive communities and in those dedicated to reproduction for religious reasons. In Western society today, however, early puberty and an average family size of three condemn a woman to perhaps 30 years of a monthly cycle culminating in slight or heavy bleeding from the uterus. The effects on her may be negligible or disabling according to the occurrence or otherwise of pain and the extent of the blood-loss, which has been estimated at anywhere between 0·7 ml and 542 ml, with an average value of 25–30 ml. Nearly one woman in five loses more than 60 ml. The heavier losses cause a marked deficiency of iron.

Like other bodily functions, the menstrual cycle is inherently variable, both between individuals and in a single individual. Quantitatively, this applies to the length of the cycle and to the length of its component parts. There is also qualitative variation in that ovulation does not always occur, in which case there is no progestational build-up of the endometrium in preparation for receiving a fertilized egg. There is also variation in the incidence of side-effects—breast changes, pre-menstrual tension, and so on.

Length of the cycle

In a large enough sample of women, the greatest number of cycles of any one length will be of 27 or 28 days, a fact which has given rise to the terms menstruation and menses. On each side of this mode the distribution of cycle lengths is fairly symmetrical between 21 and 35 days, but there is of course on each side a tail of very long cycles and very short cycles which doubtfully merit the name. Many investigations of cycle length, better known as the menstrual onset interval— from the beginning of one menstruation to the beginning of the next—have been made, and various forms of mathematics applied, but the essential facts are comparatively simple. Figures plotted (Fig. 2.5) for the distribution of 5322 cycles showed that only 1497 were of 27 or 28 days. This implies a high degree of variability and data have shown that the menstrual cycle of women can be more variable in length than the oestrous cycle of cows.[18]

For studies of this nature a large sample of cycles must of course be derived from many different women, and this will increase the apparent variability. Variation in a single individual, however, contrary to the impressions of many of those

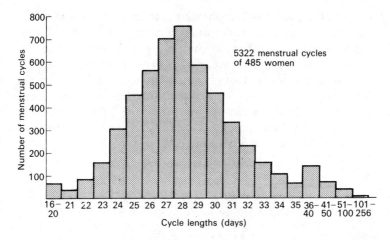

FIG. 2.5. Variation in length of the menstrual cycle in a composite sample of 5322 cycles in 485 women.[18] There are more cycles of 27 days or 28 days than of any other length, but cycles of 25, 26, 29, and 30 days occur frequently. Without more information as to their nature, it is doubtful whether cycles of less than 21 days or more than 35 days should be included in the normal range.

concerned, is often considerable. The fact was recognized half a century ago[23] when it was demonstrated not only that the length varied in one individual but that the mean length and its variability differed from individual to individual (Fig. 2.6).

Many factors besides the age of the subject have been thought to influence the length of the cycle. Emotional shock is well known to precipitate or delay menstruation, but case histories, upon which such generalizations are based, are difficult to deal with arithmetically. Of external environmental factors, climate has most often been implicated. Thus, it was reported that 192 American Army nurses transferred from temperate areas of the United States to a jungle base in New Guinea decreased their cycle length in an average of 5 months from 27·7 days to 26·9 days. The result was said to be statistically significant, but in view of the method of data collection can hardly have been meaningful. Changes in altitude have also been implicated in menstrual irregularities, but the changes in altitude concerned were far less than those experienced by passengers in jet aircraft pressurized to around 6000 ft. In the case of air stewardesses this could be an important factor, though much less so than the constant time-changes, and the consequent interference with their biological clocks, suffered by them on intercontinental routes. The results of some systematic investigation of this matter will be awaited with interest.

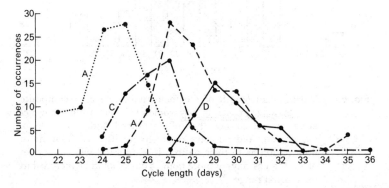

FIG. 2.6.   Individual variation in the length of the menstrual cycle.[23] In four subjects A, B, C, and D the most frequent length of cycle was 25, 27, 27, and 29 days. The variability was also different in the four subjects. Had these four records been combined an erroneous idea of variability would have been obtained.

Length of the phases and detection of ovulation

The menstrual cycle consists essentially of two phases. The first, which includes the period of menstruation, is the follicular phase, during which an egg-bearing follicle develops in the ovary and repair and proliferation of the lining of the uterus (endometrium) succeed the breakdown at menstruation and continue up to the time of ovulation, typically around mid-cycle. Ovulation is followed by the luteal phase of secretory development in the uterus, in preparation for the reception of a fertilized egg, under the influence of progesterone produced by the corpus luteum—yellow body—which develops from the ruptured follicle in the ovary. If a fertilized egg fails to appear, the corpus luteum degenerates, the endometrium breaks down, and menstruation occurs again. Unfortunately, it is not possible to forecast the time of ovulation —the vital point in the cycle—in advance, or even to detect it when it occurs, with any certainty or regularity. This fact has several implications: (a) menstruation, as the overt point of the cycle, has assumed a prominence which, as it occurs only in man, apes, and old-world monkeys, for long obscured the relation between the menstrual cycle of man and the oestrous cycle of lower mammals; (b) it makes difficult or impossible the gathering of accurate figures for the relative length of the follicular and luteal phases; (c) it complicates any method of conception control based on avoiding intercourse around the time of ovulation.

Is this situation likely to change? Can we hope that practical methods of forecasting or detecting ovulation will be developed? No doubt methods of estimating blood hormones will be further developed and we shall gain more accurate information about the relation of the changes to ovulation, but there seems little prospect of such methods becoming suitable for domestic use in the near future. Study of the biochemical and physiological correlates of ovulation also offers little immediate hope.

What remains? Some women feel a twinge of abdominal pain at about mid-cycle which has been reported to occur on opposite sides in successive months, and some have slight mid-cycle bleeding which may be associated with ovulation, but these signs, if they are signs, are far too restricted to be generally useful. One herald of approaching ovulation is much under discussion at present—the change in the consistency of the cervical mucus, which is normally scanty, tacky, and opaque, but which increases in quantity and becomes clear, smooth, and stringy under the influence of increasing oestrogen in the last few days of the follicular phase before ovulation. The volume of the pre-ovulation mucus may be sufficient to lead to an external discharge, but in any case the change is not difficult to follow. There is here the possibility of a do-it-yourself method of finding out when ovulation is likely. Unfortunately, the variability is great, and on the basis of its relation with the basal body temperature rise (see below), the stringy mucus may appear several

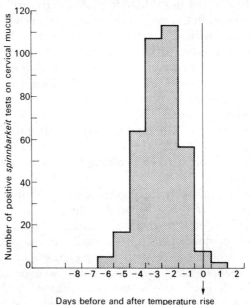

FIG. 2.7.   The changes in the cervical mucus around the time of ovulation.[13] The change to the clear stringy type of cervical mucus under the influence of oestrogen reaches a peak 2–3 days before the rise in body temperature, that is, just before or at the time of ovulation.

days before ovulation or persist for a short time afterwards (Fig. 2.7). An attempt to refine the method by the use of sensitized paper to estimate the sodium chloride content of mucus has not been of much value, because the test depends not on the concentration of sodium chloride, but on the total amount, which in turn depends on the total amount of mucus obtained.

However, in spite of the difficulty of detecting ovulation when it happens, there are methods of detecting that it has occurred a day or two previously. Thus, the presence of the corpus-luteum hormone progesterone, secreted from the ovary after ovulation, can be detected by an increase in the amount of its inactive excretion product pregnanediol in the urine. Such a biochemical method is, of course, of little value for domestic use. Progesterone, however, has the effect of slightly but detectably raising body temperature, so that a slight but consistent rise in basal body temperature (BBT), detected by daily readings, provides evidence that ovulation has occurred 1–3 days earlier (Fig. 2.8). This fact has permitted the development of the method of conception control based on avoiding intercourse until after the rise in body temperature. It is credibly recorded[37] that this method was

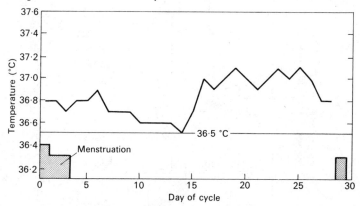

FIG. 2.8.   The rise in the basal body temperature at mid-cycle.[44] In this example the rise of about 0·5 °C took place over 2 days and probably indicated that ovulation had occurred on the previous day. Slower rises are more difficult to interpret.

originated by a biologically minded Roman Catholic village priest who knew of Van de Velde's description of the biphasic nature of a woman's BBT, and of the suggestion, as long ago as 1926, that the mid-cycle change was associated with ovulation. The priest, concerned at the repeated confessions of the use of contraceptives by his parishioners, thought up the BBT method, and dispensed advice accordingly. It seems that by 1935 he had already collected a great deal of information about the value of the method and had found that the variability of the length of the whole cycle is largely but not

exclusively caused by that of the follicular phase. This conclusion is entirely in accord with later work (see also Fig. 2.9),
but it must be remembered that all these estimates of the

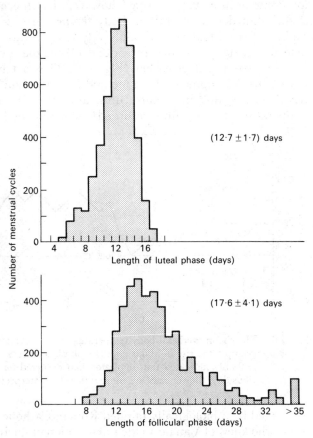

Fig. 2.9.   Relative length of the two phases of the menstrual cycle.[13]
The follicular phase is longer and more variable than the
luteal phase, but the difference in length is slightly exaggerated by the fact that these figures are derived from BBT
readings which signal late, not from the time of ovulation.

relative length of the two phases are based on the rise in BBT,
and that this may be evident at any time between 1 day and
3 days after ovulation, depending on the speed at which the

corpus luteum develops and progesterone secretion becomes effective. The variation may even be greater, because otherwise records of pregnancy resulting from a single insemination imply an unexpected survival time of spermatozoa in the female tract (Fig. 2.10). It must be noted also that the

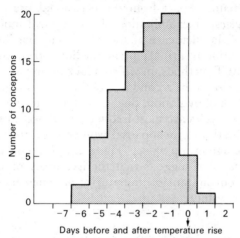

Fig. 2.10. Time of single insemination resulting in pregnancy in relation to the BBT rise.[13] Either sperm-survival is longer than currently supposed or the time between ovulation and the BBT rise is more variable.

length of the follicular phase is calculated from the beginning of menstruation and therefore includes the period of menstruation, which is the most variable phase of the whole cycle, since in an apparently normal cycle it may last for less than 2 days or more than 5 days. To what extent longer bleeding means a longer follicular phase is uncertain.

Anovular cycles, in which ovulation and the changes which normally follow it fail to occur, are not infrequent, and the pattern of their incidence is of some interest. Unfortunately, external changes indicating lack of ovulation (like those indicating ovulation) are minimal, except perhaps towards the end of the cycle, when a woman accustomed to pre-menstrual tension and breast changes might notice their absence. Failure of ovulation, however, does not result in

failure of menstruation, though the bleeding may be somewhat less; nor are anovular cycles necessarily shorter. It follows that neither the rhythm of the menstrual cycle, nor the
mechanism which precipitates the breakdown of the endometrium, is dependent on the development and regression of
a corpus luteum. Apart from the occasional use of endo-
metrial biopsies, or examination of the cervical mucus, our
knowledge of the incidence of anovular cycles has depended
on the estimation of urinary pregnanediol and on the development of BBT methods, in which a lack of pregnanediol or
temperature rise in the second half of the cycle is regarded as
indicating lack of ovulation. Such investigations have shown
that anovular cycles are most frequent in adolescents and in
pre-menopausal and menopausal women, but are least frequent in the age group 25–34 (Fig. 1.2, p. 8). This fact is of
interest, because it suggests that the frequency of anovular
cycles is a concomitant of immaturity or senescence of the
reproductive system.

Side-effects in the menstrual cycle

Basal body temperature is not the only body index to change
during the luteal phase under the influence of progesterone.
Swelling and fullness of the breasts occur not infrequently in
the second half of the cycle, to subside rapidly with the onset
of menstruation. Better known still is pre-menstrual tension,
the feeling of heaviness, pelvic congestion, and irritability
which in many women characterizes the few days before
menstruation. One fascinating study showed that criminal
offences, acute admissions to hospital, and industrial accidents were all more frequent in women pre-menstrually or
early during menstruation. Schoolgirls scored fewer examination marks pre-menstrually than at any other stage of the
cycle (Fig. 2.11). Unexpectedly, schoolgirl prefects adminis-
tered an above-average number of punishments during days
1–16 of the cycle—the greatest number during 1–4—and
few if any during the luteal phase, even pre-menstrually,
although 'offences' committed by the girls were above average at that stage. The pre-menstrual disabilities depend on

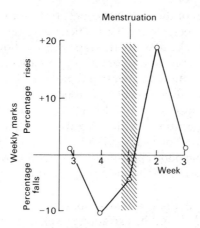

FIG. 2.11. Schoolgirls' marks in relation to menstruation.[10] The drop in the marks obtained by pre-menstrual girls is clearly shown.

the presence of a corpus luteum which in turn depends on ovulation having taken place, and are not found in anovular cycles. No doubt there will be increasing use of the pill to suppress ovulation in cases of severe pre-menstrual tension and in dysmenorrhoea as well as to fit the cycle to examination schedules. For women athletes the pill is less useful, because where regulations insist on a 2-day gap between medication and the event, bleeding is likely to appear at exactly the wrong moment.

Pregnancy

What aspects of pregnancy are likely to vary in an individual from time to time, between individuals or in different ethnic groups? Some obvious ones are length of gestation, the number of foetuses, birth-weight, and pregnancy wastage.

Length of gestation

There is little difficulty in determining the end-point of human gestation, but where can it properly be said to begin? Biologically, the starting point of gestation is evidently the fertilization of the egg, but conception can hardly be said to have taken place until the fertilized egg becomes implanted about a week later. Both these stages, however, are outwardly invisible, and in practice pregnancy mainly has to be dated from the last menstrual period (LMP). But as the operative ovulation could have occurred any time up to 3 weeks or more after the LMP, this method overstates the length of gestation. With a highly irregular cycle the overestimate may, of course, be much greater. Also to be considered is the odd report that births tend to occur not only in the early hours of the morning but also towards the end of the week.

Subject to these qualifications the length of pregnancy from the LMP and its variation is well known, and the figure of 280 days, 40 weeks, or 10 lunar months is a handy one. The size of these indices gives some credibility to the theory put forward towards the end of the last century that the length of gestation is a multiple of the length of the sexual cycle, so that in man spontaneous abortions tend to occur at 2, 3, or 4 months after the LMP and parturition can be regarded as a super-menstruation. But this idea is difficult to reconcile with the variation in the length of gestation.

Mean gestation length cited by various authorities, ranges from 270 days to 295 days; 38·6 weeks to 42·1 weeks. Overall, the minimum figure cited is 196 days (single case) and the maximum is 351 days (single case)—28·0 weeks and 50·1 weeks. Even this upper extreme is exceeded in one case recorded—of 359 days. One may question whether this extreme range is realistic, whether it results from immaturity or postmaturity, or simply results from mistakes of some sort. It has been pointed out that extremely short durations might result from bleeding early in pregnancy, thought by the woman to be a normal menstruation, while an extremely long duration might arise from an unrecognized abortion followed, without menstruation, by a durable pregnancy. Again, where a period of amenorrhoea is followed by ovulation at which conception occurs, very odd things may happen—as in a woman who has not menstruated for a year being diagnosed clinically as 5 months pregnant.

This problem is of course of legal as well as biological significance. Thus the Registrar General regards a child born more than 28 weeks after marriage as conceived in wedlock, a child born before as due to a pre-marital conception. The upper limit is also of importance in deciding how long after the death of her husband a woman's child can be regarded as legitimate. English courts appear to have accepted 349 days but rejected 360 days as possible lengths of pregnancy.

A much better estimate of the length of gestation can be made where BBT records are available for the cycle in which conception occurred. In one series of such cases, the interval between ovulation and parturition varied between 266 days and 270 days; in another between 250 days and 285 days. It is likely, therefore, that durations calculated from BBT records are not only shorter, as expected, but less variable. Cases where the duration can be calculated from a single coition are, of course, rare, though artificial insemination, which is concentrated around the expected time of ovulation, should give similar results.

There is also a relation between social class and length of pregnancy. Table 2.1 shows that although roughly one-half of all pregnancies resulting in a single legitimate birth in Aberdeen in 1951–9 were of 40–41 weeks duration, there was a higher proportion of the shorter durations in the lower social classes than in the higher ones.

For comparison with a different ethnic group, a study made in Pakistan may be considered. Between 1963 and 1965, 1013 women were interviewed at the time of childbirth. Only 28 per cent of the pregnancies were of 40–41 weeks and there was thus a much greater trend to short pregnancies than among the lower social class mothers in the Aberdeen study (Table 2.1). In the Pakistan survey, the average for single births was not affected significantly by the age of the mother, parity, sex of the child, or the income of the parents.

Another example of differences in the length of pregnancy is provided by the work summarized in Fig. 2.12, showing shorter durations in American negroes compared with American whites.[20] Other examples of what might be regarded as ethnic variation in the duration of pregnancy have

TABLE 2.1

Distribution of length of gestation in three social groups of Aberdeen first-time mothers (primiparae) 1951–9 (single legitimate births) and in Lahore, 1963–5

	Aberdeen						Lahore	
	Social class							
	I and II		III		IV and V			
Weeks	No.	%	No.	%	No.	%	No.	%
35 or less	22	1·7	169	3·0	81	4·4	71	7·1
36 and 37	40	3·1	238	4·2	86	4·7	164	16·4
38 and 39	215	16·8	1007	17·9	374	20·4	367	36·6
40 and 41	716	55·9	2913	51·8	882	48·0	285	28·5
42 or more	289	22·5	1299	23·1	414	22·5	114	11·4
Total	1282	100	5626	100	1837	100	1001	100·0

After F. E. Hytten and I. Leitch, *The physiology of human pregnancy.* Blackwell Scientific Publications, Oxford and Edinburgh (1964).

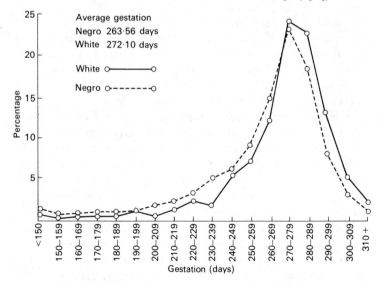

FIG. 2.12. Length of gestation in white and negro patients in the United States.[20] The peak gestation length is about 1 week less in negroes and the variability is greater.

been given, but all such records must be considered in relation to standards of living, an effect of which is suggested by the Aberdeen investigation.

Twin pregnancies are well known to be of shorter duration than single ones. In the Pakistan data, the comparative length of twin pregnancies was $(35\cdot9 \pm 2\cdot76)$ weeks and for single pregnancies was $(39\cdot40 \pm 2\cdot4)$ weeks.

Birth weight

The weight of the child at birth naturally bears some relation to the length of gestation and is influenced by prematurity or postmaturity. The accepted minimum for mature birth weight (2500 g) may be found in pregnancies ranging from 32 weeks to 52 weeks. Ethnic variation has been much discussed, but, here again, the effect of living standards must be taken into account. Not unexpectedly, for instance, the babies of 'paying' white patients in South Africa are reported as being heavier than those of the negroes. By contrast, the babies even of 'paying' patients in Calcutta and Madras are small by European standards, and there seems to be a real ethnic variation in birth weight. This, however, is not apparently necessarily related to body weight, since African pygmies, weighing perhaps 30 kg, do not have correspondingly small infants.

Secular increases in birth weight have been described, but, if real, they almost certainly derive from improvement in the standards of living. By contrast, a sex difference in birth weight is well authenticated, the newborn male averaging about ¼ kg heavier than the female.

Pregnancy wastage

Given a normally fertilized human egg, what can happen to it? It may segment normally in the female tract, become implanted normally, and proceed through normal development to a normal live birth. On the other hand, it may not. Many things can happen to it in the 38 weeks after fertilization.

It may fail at an early age through irregular segmentation or otherwise and die in the Fallopian tube or uterine cavity. Even a healthy embryo may fail to get implanted because of bad luck or defects in the uterine endometrium. After implantation it is at risk of miscarriage or spontaneous abortion because of abnormal development, genetic defect, or endocrine deficiency. The hazards are less after 4–5 months of gestation, though the foetus may still be carrying defects such as spina bifida, which have not so far proved fatal but are likely to do so later. Hazards at birth, greatly reduced in recent years, add little to the total of fertilized eggs which do not produce live-born children.

Until recently, estimates of pregnancy loss were necessarily vague. About 20 per cent of pregnancies were thought to end in abortion, but there must have been some confusion between spontaneous and self-induced abortion, only the prematurity and stillbirth rates were known with any accuracy. Modern interest, modern technology—especially chromosome technology and the advent of legalized abortion—and better medical services have altered this situation, though much is still in doubt.

Overall pregnancy loss is thought to be of the order of 50 per cent, 10–15 per cent occurring before implantation and 35–40 per cent afterwards. The latter figure is much above that given for clinically recognized abortions, usually in the range 10–15 per cent of conceptions. For example, one study in 1959 recorded that 9526 pregnancies in Belfast resulted in 8300 live births, 219 stillbirths, and 1127 abortions between the fourth and twenty-seventh weeks. The difference, of course, is accounted for by the large number of 'overdue' periods, miscarriages, and very early abortions which are not recorded clinically. None of the figures, of course, can take account of losses before implantation—which may be much above 10–15 per cent. In one study 10 out of 34 human ova recovered from Fallopian tubes were found to be abnormal. Of the causes, in another study, about 36 per cent of all clinically recognizable abortuses had chromosome anomalies, which must also account for many of the earlier embryonic deaths.

Lactation

Lactation is an essential stage of human reproduction in countries where substitutes for human milk are not available. Under these conditions failure of lactation is very serious; hence, in the more primitive parts of the world, many rites and treatments are used in trying to evoke or maintain lactation, not only in the mother but also in the grandmother or other relative. Recipes range from prayer to the drinking of plant decoctions or beating the breasts with nettles and applying hot stones. More recently it has been noted that among the low socio-economic group in India, lactating women used garlic, tamarind, and cottonseed in the belief that they stimulate the flow of milk, but no significant effect was found in a clinical test. Some of the decoctions have been examined by laboratory methods, but without clear result.

Duration of lactation

Lactation is influenced by endocrinological, neural, psychological, and social factors and is perhaps the most variable of human reproductive functions. Duration and intensity vary enormously. To this natural variation must be added the intervention of social customs, whether it be artificial curtailment of lactation, as in sophisticated areas, or prolongation to the last possible moment as in many other parts of the world, either because no other baby food is available or in the belief that lactation prevents conception.

Milk yield

The amount of milk produced by the human female may be negligible or it may compare favourably, on a body-weight basis, with that produced by a high-yielding cow. Some outstanding performances have been recorded. One woman had a peak yield of 3·5 litres per day, and her production remained high during the whole 14 months of observation. A second, though not giving quite so high a peak yield, produced during her third lactation more than 560 litres, hand-milked from her breasts, in excess of the child's requirements, a surplus which realized US $2020 at the Mothers' Milk Bureau. Even these two women appear as modest yielders compared with others recorded in the literature.

One comprehensive study of seventeen wet nurses found one woman who was able to secrete as much as 5400 ml in one day. This phenomenal record in human milk production has been surpassed more recently by a woman reported as producing 5950 ml (1¼ gallons) of milk in one day, and as able to nurse between one and seven babies throughout 51 weeks of lactation. These records are all more than 40 years old, and it would be interesting to know whether such prodigious lactations still occur. In any case, with such enormous individual variation, and with further variation caused by nutritional factors, frequency of emptying of the breast, and so on, averages have little meaning so that there are special difficulties in considering ethnic variation. An average of 500–600 ml per day for Nigerian women has been suggested, whereas a figure of 16–18 oz (about 450–510 ml) per day, rising sometimes to 24 oz (680 ml), has been given for poor Indian women.

The lactation curve

In cattle there is a sharp rise in milk yield during the first 4 weeks after parturition, and then a gradual decline during the rest of the lactation period. Figures given for Egyptian women lactating for at least a year suggest a slower rise, with peak yields at 6 and 7 months (Fig. 2.13). The mother's age made little difference to the yield or the shape of the curve, except that the decline after the seventh month was steeper in young mothers. Figures given by other workers, however, and summarized for the United States and elsewhere,[34] indicate a lactation curve more like that of the cow, in that there is a sharp rise in volume during the first week or two, which may be maintained for a time or followed quickly by a gradual decline. Whether such variations are genuine in the sense that they are due to environmental or ethnic factors or are caused by differences in the management of the lactation is uncertain.

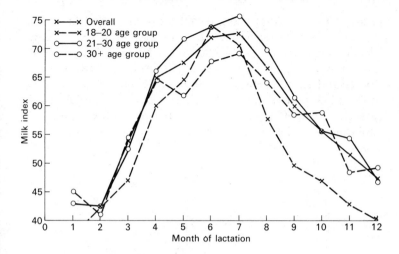

FIG. 2.13. Lactation curve for Egyptian women.[19] The relative milk yield is in units derived from test milking and not in absolute amounts. The peak yield is between 4 months and 8 months. The age of the mother has little effect under 30 years.

Composition of human milk

By happy chance, the composition of human milk is similar to that of its main substitute, cow's milk. Average figures for human milk show that water (88 per cent) and fat (3–4 per cent) are about the same; sugar (6–8 per cent) is higher and casein (about 1 per cent) lower than in cow's milk. There is considerable individual variation, both between individuals and in one individual at different times. There is also diurnal variation in the fat content which, according to observations on German women, is lowest in the early hours of the morning and highest in the afternoon. According to another study, however, the highest fat content is around 10 a.m.; it rises during emptying of the breast.

The protein content is also subject to many variables. There is even some slight difference between the left and right breasts, and considerable changes during nursing and between successive nursings. In the lactation cycle there is a big drop during the colostrum period in the first week *post-partum*, from about 12 g of protein per 100 ml to less than

2 g per 100 ml. After that, according to one set of figures,[19] there is a steady decline up to 6 months and then a slow decline, which is no doubt continued throughout prolonged lactation. Hence, where lactation is prolonged to extreme limits and supplementary food is not available or consists only of low-protein material such as bananas, the infant runs a serious risk of being protein deficient. Many studies of differences in the protein content of human milk have been

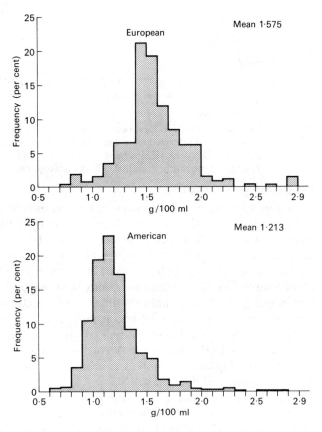

FIG. 2.14. Protein content of the milk of American and British women.[34] On the basis of these figures, variability is about the same in both countries, but English women produce milk with a higher protein content.

made. An interesting comparison between British and American women is given in Fig. 2.14, based, as the author points out, on material involving every kind of variable affecting protein content. Whether the result really implies that European women produce milk with a higher protein content than do American women or, if so, what causes the difference, is uncertain.

Fertility during lactation

If lactation does not occur after parturition, the menstrual cycle returns within 2–4 months. Some extremely interesting data relate to women in Birmingham, Manchester, London, and Newcastle in 1968. Among 93 non-lactators the mean time for the return of menstruation in the absence of lactation was about 60 days. This interval, however, was not necessarily characterized by infertility; the mean time of the first *post-partum* ovulation, as assessed by BBT measurements, was more than 70 days, but one-third of the patients ovulated, and were, therefore, potentially fertile before the first *post-partum* menstruation. This could mean two births in less than a year to a woman who assumed that *post-partum* amenorrhoea implied infertility, a situation reminiscent of that at puberty (see p. 18), and one which should be emphasized by family planning organizations. By contrast, among 81 lactators it was found that the interval between childbirth and the first menstruation and ovulation increased, as might be expected, with the duration of lactation. The figures implied that many women did not menstruate or ovulate before the end of lactation, and showed that almost one-half of the lactators ovulated before the first *post-partum* menstruation, but that provided menstruation had not returned, ovulation before the end of week 10 *post-partum* was a rare event.

Strict comparison of these figures with those for other parts of the world is not easy, but the following give some indication of possible differences. In South India, it was found that only one-third of lactating mothers resumed menstruation within 5–8 months of delivery and little more than one-half before 18 months. Such differences from the British figures are presumably due to differing standards of nutrition and the prolongation of lactation. In the Central African state of Rwanda, a study showed that one-half of the non-lactating mothers conceived again within 4 months but in lactators the corresponding figure was 18 months (Fig. 2.15).

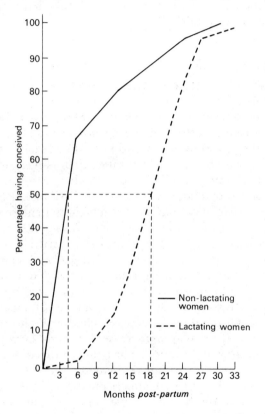

FIG. 2.15. Time of conception *post-partum* in lactating and non-lactating Rwanda women.[3] One-half of the non-lactating women were pregnant again within a few months *post-partum*; lactating women took more than a year longer.

3

Seasonal changes in
sexual activity and the birth-rate

Many mammals breed only during a restricted period of the
year, the breeding season. In the off-season, reproductive
function, physiological and ethnological, is in abeyance in the
female and usually in the male. This restriction may be of
genetic origin, so that, as in some equatorial animals, it
occurs in an environment in which there is no obvious seasonal change, or it seems to be independent of environmental
changes. In either case, the cycle is presumably regulated by
a built-in biological clock. More frequently, the incidence of
the breeding season is determined by a combination of genetic and environmental factors, in that the animal has a
genetic capacity to adjust the breeding season to the environment. In such cases the breeding season, though sharply
defined, is susceptible to external influences, notably light.
The effect of external factors, however, is not necessarily the
same in different species. In the ferret the breeding season is
associated with the increasing light of spring, in the sheep
with the waning light of late summer and, in both, the breeding season shifts appropriately on transference from the
northern to the southern hemisphere, or under conditions of
artificially changed lighting.

Again, an animal may have the capacity to breed at all
seasons of the year, but show some reduction at certain times
depending on environmental, nutritional, or other conditions;
examples of this type of seasonal breeding are to be found
among many animals. It has long been known that in Rhesus
monkeys, under laboratory conditions, there is a concentration of conceptions within the last quarter of the year in the
northern hemisphere and a tendency to reverse this incidence

when the monkeys are transferred to the southern hemi-sphere.[18] The periods of reduced breeding appear to be associated with an increase in anovular cycles.

What relevance, if any, does this have for man? Is there any evidence, for instance, that length of daylight has any influence on the incidence of human reproduction?

Length of daylight

There have, of course, been many reports of restricted breed-ing seasons in some communities in different parts of the world, often ascribed to changes in the length of daylight. The most persistent and most plausible accounts relate to the Arctic regions, where the alternation of 6 months' daylight and 6 months' darkness might be expected, by analogy with animals, to exert a maximal influence. Thus, Dr. Cook, ethnologist to the first Peary North Greenland expedition, writing of the Eskimos there, said: 'During the long Arctic night the menstrual function is usually suppressed, not more than one woman in ten menstruating.' And again: 'During the whole of this long Arctic night the secretions are dimi-nished and the passions suppressed, resulting in great mus-cular debility.' Some subsequent observers tended to confirm this report, but contrary ones are not lacking; for instance, a Danish Governor of a colony in Northern Greenland married a full-blooded Eskimo woman, raised a family and, as a result of his first-hand information, reported that there was no apparent decrease in menstruation or sexual desire dur-ing the winter months. More factually, an analysis of the incidence of 16 101 births in West Greenland between 1901 and 1930 indicated that conceptions took place freely through-out the year, but there were two small peaks, in December during the Arctic night and in April on the return of the sun.

It is very doubtful, therefore, whether there is any evidence for any part of the world of a sharply defined and restricted human breeding season, conditioned by the length of day-light or otherwise. Nevertheless, there is in many countries seasonal variation in the birth-rate to the extent of a differ-ence of 10–15 per cent between the peaks and the troughs. A

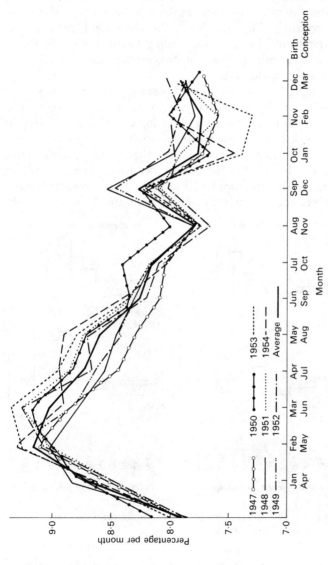

FIG. 3.1. Birth-rate and conception-rate in Bavaria, 1947–54.[26] The situation appears to be identical with that in England and Wales.

typical example for Europe is given by a study of seasonal variation in the birth-rate in Bavaria for each of the years 1947 to 1954. Fig. 3.1 shows the clear peak of births in February and March corresponding to conceptions in June–July and a minor peak in September corresponding to conceptions in December.

It might be thought that the main peak was associated causally with increased day-length in summer. Credence is given to this idea by the fact that the birth pattern, like that of day-length, is reversed in the southern hemisphere, especially in New Zealand and South Africa where, irrespective of race, the peak conceptions occur in December and January—also in summer (Fig. 3.2). This attractive idea, however, receives a rude shock from the fact that in the

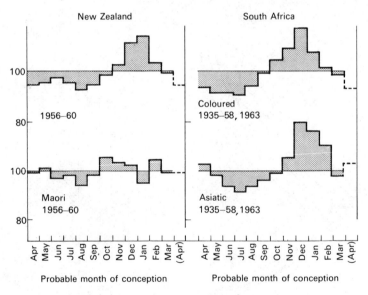

Fig. 3.2. Seasonal variation in the conception-rate in New Zealand and South Africa.

United States, in spite of its position well north of the equator, the variation in the seasonal birth-rate, calendar-wise, is similar to that in the southern hemisphere, and opposite to that in Europe (Fig. 3.3). The variations thus appear to be of social rather than physiological origin. In support of this interpretation it may be noted that, over a period of 20 years,

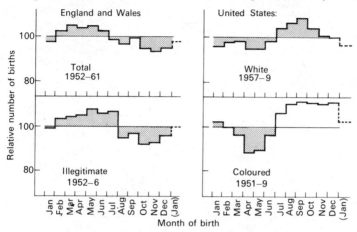

FIG. 3.3.   Birth-rate in England and Wales and in the United States.[8]
The seasonal patterns are reversed.

the seasonal variation in the birth-rate in Puerto Rico (Fig. 3.4), presumably as a result of American cultural influences, changed from the European to the American pattern.[2]

The comfort factor

The extent, if any, to which climatic factors other than the length of daylight may contribute to seasonality in the birth-rate in temperate climates is uncertain, but there is good evidence of what may be called a 'comfort factor' in sexual behaviour. Few people, one imagines, would seek intercourse at an ambient temperature of, say, 45 °C and a humidity of 95 per cent. Without thinking of such extremes, there are some seasonal variations in birth-rate, and thence presumably in conception-rate, which could well be attributed to the seasonal appearance of climatic conditions discouraging to coitus. Some work has suggested that the conception-rate is highest when outdoor temperatures range around 18 °C and much reduced when the mean temperature goes above

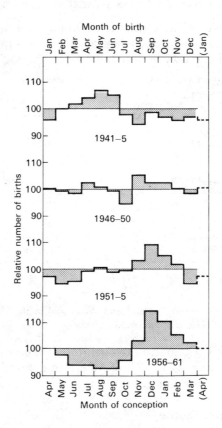

FIG. 3.4.   Birth-rate in Puerto Rico 1941–61. The pattern changed from the European to the American pattern over 20 years.

21 °C. In Montreal, where summer warmth is never oppressive, the conception-rate hits a peak in mid-summer and is lowest in the cold winter; in Cincinatti there is a definite drop in the conception-rate during the hot summer; and at Tampa in Florida conceptions are fully a third less during the hot humid summer than during the mild cool winter. All through the American Middle West there was a sharp drop in conceptions during the blazing summer of 1934. It would be interesting to know what effect the spread of air-conditioning has had on such fluctuations.

Clear-cut seasonal variations are found in the conceptions in Hong Kong, there being a peak in winter and a trough about 30 per cent lower in summer (Fig. 3.5). Various possible explanations, including social ones, have been considered; for instance, it has been pointed out that the high conception-rate and low temperature in January correspond with the Chinese New Year, so that social effects cannot entirely be excluded. The conclusion, however, is that temperature is the operative factor, fertility being maximal at around 15 °C and minimal at 27 °C and higher. Whatever its cause, the variation is consistent from year to year, even though the birth-rate has declined steadily.

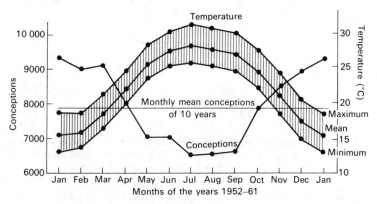

FIG. 3.5.   Conception-rate and air temperature in Hong Kong.[6] There is a clear inverse relation.

In a study of the effects of temperature on human physiology, the seasonal variation in the birth-rate in Tasmania was compared with that in other parts of Australia.[30] The two extremes are shown in Fig. 3.6. Considering that inland Queensland is stressfully hot in summer and Tasmania is unpleasantly cold and wet in winter it would appear that in each case the conception-peak occurs at the most equable time of year, the winter in inland Queensland and the summer in Tasmania. The seasonal variations are not very great, being, in terms of an annual mean calculated as 1000, from

900 to 1050 in the first case and from 950 to 1100 in the second, but the trends are unmistakable.

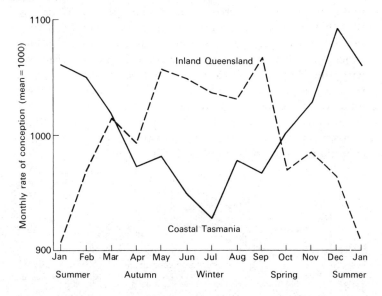

FIG. 3.6.   Conception-rate in coastal Tasmania and inland Queensland during two decades.[30] In each area the peak number of conceptions occurs at the most equable time of year.

Equally striking are the figures for Singapore passed to me by Wing Commander J. H. Rogers. Among other indices showing seasonal variation, Rogers collected figures for birth-rate in relation to 'effective temperature', an index which is reminiscent of the American 'comfort factor' and which for the optimal comfort of Europeans in Singapore was found to be 24·4–25 °C. The records indicate that during the hot season the average monthly effective temperature is mainly above the optimal for comfort, implying stressful conditions on many days. There is here a remarkable inverse correlation between effective temperature and conception-rate, the conception-rate over 5 years having a correlation of −0·64 with the effective temperature (Fig. 3.7). It is probably significant that the comparative failure of the temperature to drop in the middle of 1963 is associated with the

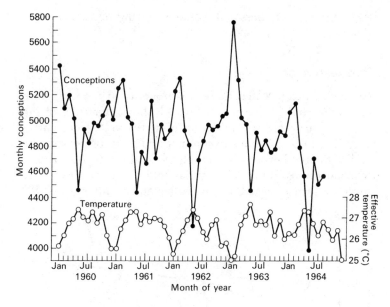

Fig. 3.7. Conception-rate and effective temperature in Singapore over 5 years (from the data of Wing Commander Rogers). There is a close inverse relationship.

smallest peak in conception-rate and the biggest subsequent drop over the 5 years. As found by other authors,[30] the differences between the peaks and troughs of the conception-rate are not great (less than 25 per cent), but appear constant and meaningful. The regular variation was shown most clearly by the large group of Chinese, to a lesser extent by the smaller group of Malays, and not at all by the minority of Indians, Pakistani, and other races. Rogers considered, in some detail, possible effects of the Chinese New Year and the Muslim Ramadan and came to the conclusion that the essential factor in the seasonal variation is climatic rather than cultural. There is little need to suppose that the decreased conception-rate during the hot season is due to physiological damage to the reproductive organs, and it would seem reasonable to ascribe it to a lower coital rate resulting from heat fatigue and distaste for personal contact during extremely hot weather.

The festive seasons

In England and Wales there is seasonality in the birth-rate similar to that in Germany. Fig. 3.8 shows, for each of the 4 years illustrated, a high peak of births in March, corresponding to conceptions in summer and a minor peak at the end of September and beginning of October, corresponding to conceptions around Christmas and the New Year. The same pattern is seen both in earlier and in later years. To

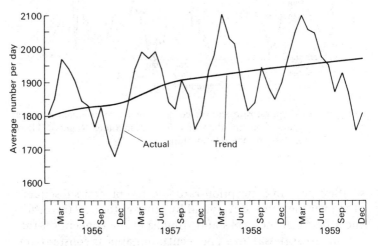

FIG. 3.8.   Legitimate birth-rate in England and Wales, 1956–9.[11] The major peak corresponds to conceptions in the summer and the minor one to conceptions around Christmas and the New Year.

what are we to ascribe these peaks and valleys? Length of daylight, as we have seen, does not appear to be involved in variation of this kind, and climatic conditions in England and Wales are rarely such as to discourage coitus. What remains?

Is there any evidence of seasonal variation in pregnancy wastage which might cause seasonality in the birth-rate, and upset calculations of the conception-rate? Current knowledge does not permit an answer, but the possibility should not be overlooked, especially as there is some evidence from the United States of seasonal variation in the conception of malformed children. On present knowledge, however, it seems that seasonal variation in the birth-rate is due primarily to variation in the conception-rate, and that we have, therefore, to explain a slight but regular seasonal variation in the conception-rate. Could this be due to changes in female fertility? It is doubtful whether the overt established menstrual cycle shows any changes that might suggest such seasonality, but one cannot entirely exclude the possibility of seasonal variation in the incidence of anovular cycles, such as was found to be associated with the seasonal variation in the conception-rate in Rhesus monkeys. A case has, in fact, been recorded of a 24-year-old woman who showed what appeared to be a seasonal grouping of anovular cycles as judged by basal temperature records, but the concentration was in the summer, not as might have been expected in the winter.

What about the male? In the absence of evidence of any seasonality in the vigour of sperm production, such as might show up in sperm counts, is there any evidence of variation, for psychological or other reasons, in male sex drive, leading perhaps to seasonal variation in the frequency of intercourse and thence, probably, to variation in the conception-rate? Direct information may well be lacking on so difficult a question, but there may be indirect information. In 1892 Alfred Leffingwell[25] wrote a most interesting little book entitled *Illegitimacy and the influence of seasons on conduct.** Among a lot of illuminating material, Leffingwell had a diagram of the seasonal incidence of what he euphemistically called 'offences against chastity' (Fig. 3.9). The concentration of such offences in the second and third quarters of the year is perhaps suggestive of seasonal variation in sex drive, but it

* I acquired this book at a second-hand bookshop in Manchester for ninepence in 1922. It had originally been published, together with fifty or so other volumes in a Social Science Series. I mention this to show that the idea of social science is far from new, so that even three-quarters of a century ago a relatively obscure publisher could produce fifty volumes on it.

could also be merely the result of increased opportunity. In any case, the implications are not straightforward, because another diagram given by Leffingwell shows a similar seasonal variation in the suicide rate, as do more modern

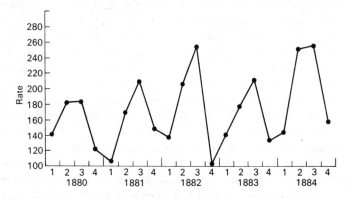

FIG. 3.9. Number of crimes against chastity in England and Wales 1880–4.[25] The summer peaks could imply greater sexual activity or greater opportunity at that season of the year.

data. Moreover, seasonality in male sex drive could produce seasonal variation in the birth-rate only if the male determines, for the most part, the frequency and timing of intercourse, and if there is a relation between the intercourse-rate and the birth-rate. Both of these propositions were probably true in the days of Leffingwell, but times have changed. Modern contraception, however, has thrown up a most curious piece of information bearing on this problem. The Marie Stopes Memorial Centre has a considerable mail-order and over-the-counter business in contraceptives, mainly in what used to be known, before the Pill became respectable, as 'conventional types'. These sales, in 1963–6, showed a distinct seasonality, and one would expect such variation to be correlated inversely with the conception-rate. In fact, the correlation is a positive one (Fig. 3.10). This unexpected result, which has since been confirmed, seems inevitably to imply a much increased frequency of intercourse during the summer. It also poses a question. Is there a difference in seasonality in planned and unplanned conceptions or in the demand for legal abortion? An answer to this question could be of interest. In the meantime, we should ask whether the seasonal variation in the frequency of intercourse, suggested by the seasonality in contraception sales, is likely to be of biological or cultural origin.

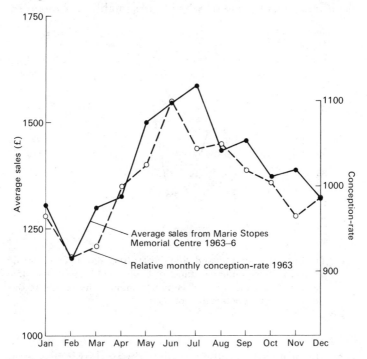

FIG. 3.10. Conception-rate and sales of contraceptives.[35] The correspondence of the summer peaks implies greater sexual activity at that time.

In explanation of the spring peak in the birth-rate it has been argued, in the context of the mid-twentieth century, that people prefer to have their babies in the spring when the nights are getting shorter and the mornings lighter, and/or that to get the maximum tax rebate for the minimum of expense people arrange to have their babies near the end of the fiscal year (5 April). Either theory, however, implies that the seasonal variation is the result of family planning, and there are two good reasons for rejecting this implication. First, the same thing occurred a century ago, before the days of heavy income tax and when family planning, if any, was rudimentary (Fig. 3.11). Second, the same cycle is seen with illegitimate births, few of which, presumably, are planned

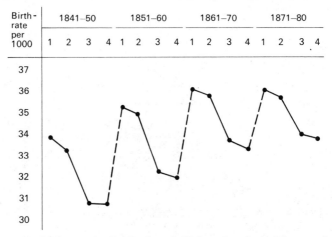

FIG. 3.11. Birth-rate in England and Wales 1841–80.[25] These spring peaks, very similar to those of today, could not at that time have been the result of family planning. The birth-rate was more than double that of today.

for domestic convenience, fiscal relief, or any other reason. In 1939 the seasonal variation in illegitimate births was even greater than in legitimate ones (Fig. 3.12), which is what one might expect if the majority of illegitimate conceptions took place under what Marie Stopes euphemistically referred to as 'alfresco' conditions. This difference has now disappeared, which in itself is a significant comment on social changes, but seasonal variation still continues in both (Fig. 3.13).

If, therefore, family planning is not the explanation of the annual cycle in the birth-rate, what is? There are plenty of clues to the answer. The larger summer peak in conceptions and the smaller winter one are reminiscent of the one-time

mid-summer and mid-winter festivals celebrating the longest
and shortest days of the year, festivals which now take the
form of the summer holidays and the concentrated festivities

FIG. 3.12. Legitimate and illegitimate birth-rates, England and Wales,
1939. Note the greater variation in illegitimate birth-rate.

of Christmas and the New Year. After all this, therefore, we
arrive at the platitudinous conclusion that conceptions are
most frequent when people are most festive and carefree.

At one time I thought I had a beautiful example of the
reverse of this picture. In 1957 the usual sharp peak of spring
births corresponding to conceptions in the summer of 1956
was truncated (Fig. 3.1). What happened in the summer of
1956 to decrease the conception-rate? The obvious answer
is the beginning of the Suez crisis—when many people were
worried about the possibility of another war.

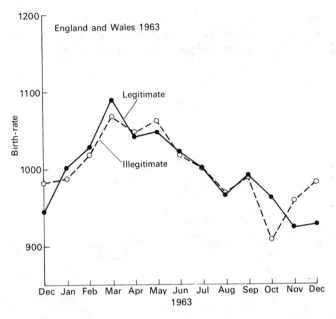

FIG. 3.13. Legitimate and illegitimate birth-rates, England and Wales, 1963. The disappearance of the difference in the patterns implies increased opportunity for extra-marital intercourse.

However, further investigation showed that in the 10 years 1952–61 three other spring birth-peaks, in 1953, 1954, and 1960 had had flat tops. From consideration of the general trend, it would seem that these flat peaks are more likely to be the result of expansion rather than of truncation, but in either case I have yet to discover what crises happened in the summers of 1952, 1953, and 1959 to cut the conception peaks, or alternatively what sources of euphoria arose to extend them sideways. Some, at least, of those who read this book must have contributed to these flat peaks. Information please!

4

Environmental influences on fertility

Man's environment is made up of the physical and nutritional conditions and the social and cultural influences under which he lives. The direct effect of physical conditions on the organs concerned with reproduction is probably negligible. Indirect effects, however, especially on sexual activity, can be important (see p. 51). Even the effects of malnutrition may be indirect rather than direct on the reproductive organs.

In contrast, human fertility in the demographic sense of reproductive performance is much affected by social and cultural influences such as religious indoctrination, local *mores*, economic conditions, technological development, and so on.

The physical environment

Temperature

In man and most other mammals spermatogenesis is dependent on the testes being maintained at a temperature lower than that of the body as a whole and to this end the testes in the adult are not only sited outside the abdominal cavity, but are carried in a pendulous scrotum which by relaxing to diffuse or contracting to conserve heat exerts an essential thermo-regulatory function. Interference with this mechanism, either by excessively high ambient temperatures or by scrotal insulation, depresses spermatogenesis to a greater or lesser extent. It may be that climatic conditions in some parts of the world provide sufficiently high temperatures to produce this effect, but allowing for the cooling effects of sweating, it is doubtful whether infertility of this kind is caused directly in man by atmospheric temperatures compatible with otherwise normal functioning of the body.

The effect, however, is a real one and other forms of ambient temperature are more effective. This fact, 50 years ago, prompted warnings about the possible effects of hot baths on male fertility and it is said that at the height of the agitation the only thermometer available disappeared from a famous Department of Zoology.

As to scrotal insulation, heavy underwear has been named as a hazard, and the suggestion made that its use by men in winter may be a factor in the lower conception-rate at that season of the year. Hard evidence on the underwear story is lacking, but attempts have been made, allegedly with some positive results, to reduce male fertility in man, as is possible in domestic animals, by enclosing the scrotum in a bag made of highly insulating material. It may be, however, that the human testis, after centuries of some degree of insulation is losing its sensitivity to deep body temperatures, a suggestion for which there is some experimental as well as other evidence.*

Altitude

Experiments have shown that damage to the testes, resulting in defective spermatogenesis, is produced in laboratory animals by maintenance under conditions of reduced atmospheric pressure. The operative factor appears to be reduced oxygen pressure. It is not surprising, therefore, that there are numerous reports of domestic animals becoming sub-fertile or infertile at altitudes above 3050 m (10 000 ft). In the Andes, such reports started in the early days of the Spanish conquest. Thus in A.D. 1535, Pizarro transferred the capital

---

* At a well-known restaurant in Delhi the cooks, clad only in loin cloths, squat on top of 'tandoors', cauldron-shaped ovens with the open top at floor level, half-full of red-hot charcoal. The possibilities are obvious to any reproductive biologist, one of whom watching the cooking process, said to me 'I'm going to talk to one of these fellows'; the following conversation ensued: 'That was a good tandoori chicken you did for us. Do you like the work?' 'Oh, yes!' 'Do you have children?' 'Oh, yes'—'How many?' 'A large family'—'Have you been doing this sort of work for a long time?' 'Nearly all my life, like my father before me, and his father before him.' And that was that.

of Peru from Jauja at 3300 m (10 900 ft) to Lima at sea-level because, as specifically noted in the Act founding the new capital, horses, pigs, and fowl would not breed at the high altitude. Such reports have been abundantly confirmed in modern times by records of infertility in horses, cattle, and sheep at great heights, and a note that sheep showed a 75 per cent lambing rate at 3500 m (11 500 ft), but scarcely one-half that rate at 4500 m (14 750 ft).

These observations could obviously have some significance for man, especially as it has been estimated that at least 10 million people around the world live at altitudes between 3600 m and 4000 m (11 850–13 100 ft), most of them in the Andes. In Peru, the vertical distribution of the population as described by Monge[33] is remarkable, about one-half of the population living at sea-level (Fig. 4.1) and most of the other half at heights between 2000 and 4000 m (6600–13 100 ft). An even more extreme distribution presumably exists in Bolivia, which has no coastal plain and of which the capital, La Paz, is at an altitude of some 3500 m (11 500 ft). Unfortunately, although high-altitude diseases and adaptations of man have been described in detail, the effect of altitude on reproductive potential is inadequately studied. Hence the prominence given to accounts of fertility, or lack of it, in the early Spanish settlements in the Andes. Thus, an ancient account has it that the first child to survive among the Spanish colonists in Potosi at 4500 m (14 750 ft), now in Bolivia, was born (presumably to a later settler!) 53 years after the founding of the city in the sixteenth century, although the fertility of the natives was unimpaired. The child in question subsequently proved his own fertility by having six children, all of whom died in infancy. Ultimately, the fertility of the Spaniards was restored, partly at least by outbreeding with the Incas.

More recently, the International Physiological High Altitude Expedition of 1935 visited the mining camp at 5300 m (17 450 ft) on Mount Aucanquilcha in Northern Chile, probably the highest permanently inhabited place in the world. The men were unable or unwilling to live nearer the mine at 5600 m (18 400 ft). The expedition noted that the women and children in the camp fared as well as the men. The women went down to Ollagüe, 3600 m (11 850 ft), to give birth to their children, but returned a few weeks later. Presumably, therefore, fecundity is not extinguished in acclimatized men and women at 5300 m. This observation accords with those of Monge, who concluded that human reproduction is possible up to any altitude at which man can live permanently.

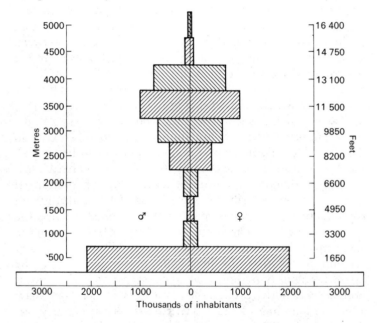

FIG. 4.1.   Population of Peru, grouped by districts, according to the altitude of the district's capital, showing diphasic distribution of males and females.[33] About one-third of the population lives at 3000 m or above.

Light

The evidence that light plays a large part in regulating the breeding season in many animals is conclusive, but for man the evidence is equivocal, if not definitely negative. One survey found a marked seasonal variation in the onset of menarche, many more girls experiencing their first menstruation in the summer than in the winter. The rise and fall of the curve follows about 2 months behind that for the length of daylight, but there is no evidence of a causal relationship. Moreover, it appears that blind girls experience menarche at about the usual age or even slightly earlier. Thus, in one series recorded, 85 prematurely born girls, blinded following exposure to excessive amounts of oxygen, and without light perception, experienced menarche at an average of 143 months, 7 months earlier than 98 prematurely born girls with normal vision.

The lack of correlation between annual daylight cycles and the seasonal variation in the birth-rate in man has been discussed in Chapter 3. Seasonal variation in the onset of menarche has also been ascribed to effects of light. The curve given by Valsik shows a peak incidence of menarche in August and September each with about 15 per cent of the total, and a trough in March, with only about 4 per cent. The curve for the incidences of menarche thus follows 2–3 per cent behind the light cycle but there is no evidence that the effect is a causal one.

The nutritional environment

In his hunter–gatherer and early agricultural days man's food supply could properly be regarded as part of his environment and the same is still true in remote parts of the world, where the people still live on local products. In the market at Iquitos, on the upper Amazon, as I have seen, foodstuffs include smoked monkey, giant snails, dried fish, and small mammals, together with various fruits and roots, all obtained from the surrounding jungle and river. A mixed diet of this kind, if sufficiently plentiful, is not likely to be inadequate for human reproduction. In other parts of the world reliance, for cultural or economic reasons, on a single crop may lead to difficulty, especially when that crop fails. Better distribution of the world's food is gradually breaking down this kind of isolation, but the recurrent outbreaks of famine in various parts of the world is a stern reminder that the problem is not yet solved.

The effect on birth-rate, however, should not be exaggerated. Man can and does breed under conditions of severe undernourishment, as witness the occurrence of pregnancy in Nazi concentration camps and of a high birth-rate in the near-famine areas of the world. The foetus, being parasitic on the mother, has priority for current food intake and for body stores, as, for instance, in the special case of the decalcification of the pregnant mother's teeth and bones, where the diet is deficient in calcium. The effects of malnutrition in affecting population growth have depended more on paving the way for disease, especially in infants, than on decreasing the birth-rate. Low birth weight arising from malnutrition, for instance, has often been described, but here we must restrict ourselves to direct or indirect effects on the gonads.

Apart from the general level of food intake there is the qualitative aspect. And here we are largely ignorant of the effect of qualitative deficiencies on the human reproductive organs, for the obvious reason that observations are limited and often equivocal and that few people would wish to subject men and women to the rigours of exact experimentation. There are three essential constituents of a complete diet: vitamins; certain amino acids, which determine protein quality; and trace elements. Qualitative malnutrition has occurred mainly where diet is restricted to one or two items, the best-known case being where an exclusive diet of polished rice from which vitamin B_1 had been removed with the husk led to the condition of beri-beri. It is not known, however, whether this had any effect on reproduction apart from that of severe debility. A related disease, pellagra, caused by lack of other components of the B vitamin group occurs where the staple diet is maize, but again specific effects on the reproductive organs, if any, are unknown.

Much the same probably applies to another well-known vitamin deficiency disease, scurvy, caused by lack of vitamin C, a disease much dreaded by sailors and explorers before it became known that it could be prevented by citrus juices, rich in vitamin C. Again, rickets caused by absence of vitamin D and common in industrial towns a hundred years ago does not seem to have depressed the birth-rate in those populous areas.

Of the other two well-known vitamins, the fat-soluble A and E, A occurs characteristically in milk and fish oil and deficiency causes night-blindness and other eye abnormalities and lack of resistance to infection; E, found especially in wheat germ oil, is known to be necessary for the completion of pregnancy in rats, and there has been some suggestion that this may apply to women. The matter is not important because an E-deficient human diet must be a rarity.

The essential amino acids may be assumed to be necessary for spermatogenesis, in view of the high demand of that process for protein.

Of the trace elements, zinc has been implicated in spermatogenesis in rats, in which its inhibition by cadmium causes breakdown of the testis, and zinc deficiency in man has been recorded as delaying the processes of puberty. Curious effects on the sex ratio have been reported to follow variation in the amount of some of the other metals in the drinking water.

The effect of malnutrition on reproduction and the reproductive organs is not necessarily direct. Severe protein deficiency is known to cause liver damage, so that the liver is unable to carry out its usual function of inactivating oestrogen. In the male, in which oestrogens are unexpectedly abundant, possibly as by-products, liver damage can therefore result in an accumulation of oestrogen. This in turn leads to effects on the mammary glands and to inhibition of the activity of the anterior pituitary gland. As a result, gynaecomastia (development of the male breasts) and atrophy of the testes may appear. Effects of this kind were noted after rehabilitation in prisoners of war returned from Japanese prison camps and more particularly have been described in Africans with a history of kwashiorkor,* a protein deficiency disease common among African children and likely to cause permanent liver damage, and therefore failure to inactivate oestrogens in the male. A well-known pathologist working in Nairobi

* An African word meaning 'the disease which comes after the next baby is born' (the elder child being weaned on to protein-deficient foods).

described to me, as follows, features of the African male which could be ascribed to this condition:

> Our own experiences and the records of others have drawn our attention to the feminine appearance of many African males, to the high incidence of gynaecomastia, to the fact that one-fifth of the breast carcinoma in Africans in Kenya occurs in males, to the local difficulty in sexing bones, to the trouble the Army had with Africans due to their development of mastitis under webbing equipment, and to the frequency of testicular atrophy.

The social and cultural environment

The trend towards small families

The birth-rate in every European country fell drastically between 1880 and 1930, though not synchronously or at the same speed (Fig. 4.2). There is no evidence that the decline was due to reduction, physiological or behavioural in origin, in the potential fertility of Europeans; presumably, therefore, it was due to birth-control by some means or other, supple-

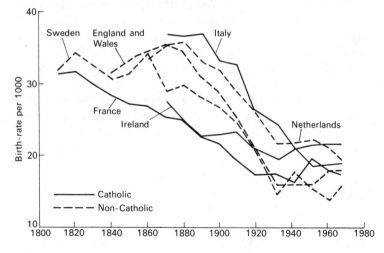

FIG. 4.2.   Birth-rates in certain European countries over 150 years.[39] There were big differences in the rates of decline, extreme rates being shown by France and Italy.

mented by later marriage and motivated by a fashion for smaller families. In these circumstances, it is curious that the decline both in the birth-rate and in family size should have affected the predominantly Catholic countries equally with the predominantly Protestant ones.[39]

In France, the fall in the birth-rate started very early, even in the seventeenth century, and from 1810 to just before the First World War it fell from 32/1000 to 20/1000. By 1930 it was down to 15/1000, at which point it was joined by that for England and Wales where the fall had started later in association with a fall in family size from more than 6 in the 1860s to slightly above 2 in the 1930s (Fig. 4.3). In Italy, the fall in the birth-rate was exceptionally rapid, so that, although it started much later than in other countries, the birth-rate

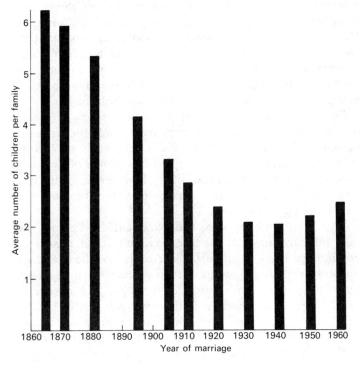

FIG. 4.3.   Decline of family size in England and Wales, 1860–1960.[41]

reached nearly the same low point by 1940. One may well wonder how, in Catholic countries, before the advent of the rhythm method, the social trend towards small families was reconciled, intellectually and contraceptively, with the cultural prohibition of birth-control.

Retention of high fertility

The trend to smaller families and a lower birth-rate, which started in most European countries a century or more ago, was prompted at least in part by the decrease in child mortality, which meant that it was not necessary to have a large number of children to ensure that some survived. In the East, and in Central and South America the fall in infant mortality came later but much faster, and in those areas social *mores* have not yet caught up with the new situation. Hence the tremendous growth of population. No doubt the current propaganda and development of family-planning services will bring the situation under control in time, but even so the idea that a large family adds to a man's prestige, and provides for assistance in his old age, will die hard.

Of more immediate interest is the occurrence of pockets of culturally based high fertility in areas where the birth-rate in general has fallen substantially. Several such pockets still exist in the United States, occupied by colonies of two small Protestant sects with outstandingly high fertility, the Hutterites and the Amish, both derived from the Anabaptists of sixteenth-century Europe.

The Hutterites are theocratic and anti-intellectual, but take full advantage of modern techniques and regard the community as the social unit. The Amish, also theocratic and anti-intellectual, eschew modern inventions such as electric light, telephones, and motorized vehicles, and regard the family as the unit of society. Both sects are dedicated to farming and reproduction, and apply to the limit the biblical injunction to be fruitful and multiply. Their demographic indices are remarkable for modern America or, for that matter, for anywhere in the West today. In the years 1941–50, the Hutterite birth-rate was 45·9/1000; and an Amish who died in 1961, shortly before his 95th birthday, left 410 living descendants. Hutterite women commonly produce children when 40–50 years old (Fig. 4.4): of the 201 Hutterite women surveyed in one particular study,[40] 139 had had their final conception between these ages. In the early 1950s, the average completed family size was over 10, which leaves far

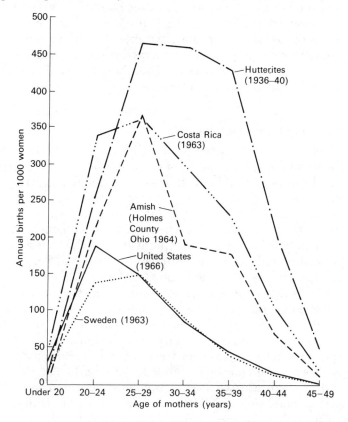

FIG. 4.4.   Age-specific birth-rates for Hutterite and other populations.[40] The Hutterites have by far the highest rate, especially for mothers over 30 years of age.

behind even the best efforts of the nineteenth century Mormons, for whom sibships containing one set of twins ranged from 2 to 19, with an average of 7·67. In the 1950s also, more than 50 per cent of the Hutterite population was under 15 years of age which, if not a world record, is remarkable for contemporary North America. During the present century the population of the United States has increased by between 1 per cent and 2 per cent per year; the Hutterite increase has been 4 per cent to 5 per cent per year. Were it not for the facts that pre-marital intercourse is banned and that marriage is comparatively late among the Hutterites (being necessarily preceded by adult baptism), the growth of their numbers would be even greater. It is perhaps fortunate for the rest of the country that the American Hutterites numbered only 440 when they first established colonies in the United States.

Marriage and coital patterns

Marriage patterns range from celibacy to monogamy and thence to polygamy, and from early marriage to late marriage. Little need be said about celibacy in relation to fertility, except to marvel at the effrontery of celibates who, themselves practising the ultimate form of birth-control, attempt to ban more practical methods for other people. Strict monogamy is compatible with maximal reproductive performance of a population only when the numbers of males and females of reproductive age are exactly equal. Polygamy, in the sense of polygyny, has sometimes been blamed, in the East, as a factor in the population explosion but, unless there is something unusual about the sex ratio, this must be fallacious; if one man has four wives presumably three men have no wives. Polygyny may even decrease the overall birth-rate, since it should decrease the frequency of coitus for any particular woman and is often associated with prohibition of intercourse for long periods after childbirth. In this connection, it may be recalled that, in the absence of lactation, fertility returns rapidly after parturition and that lactation, even when prolonged to the last moment with that aim, does not give indefinite protection against a new pregnancy (see p. 41). A fashionable decline in breast-feeding will increase fertility. Early marriage, of course, will have a similar effect, since young women are more fertile than older ones, coital rates are higher in younger couples, and exposure to the risk of pregnancy is potentially longer.

Coital patterns also become important where there is cultural prohibition of intercourse at some time of the cycle. The best-known instance of this is to be seen among orthodox Jews, for whom intercourse is forbidden for 7 days after the end of menstruation. This has the effect of delaying coitus to the fertile phase of the cycle, which, other things being equal, should materially increase the chance of pregnancy. One may speculate as to whether the ancient Jewish custom was based more on some intuitive understanding of physiology than on cultural considerations. Conversely, is it possible, in the light of the population explosion and the small family trend, that the essential feature of the strict rhythm method of birth-control, the restriction of coitus to the second phase of the cycle, will one day attain among Catholics the dignity of a religious observance rather than a contraceptive procedure?

Effect of twins on family size and birth interval

Young children are part of the parents' environment—the most important part for the mother. What effect do the early ones have on the future reproductive performance of the parents; especially, what is the effect of twins, which may be said to constitute a super-environment? In a survey of a large number of modern Swedish families containing twins, there was a significant difference in the number of subsequent siblings according to the sex of the twins. Thus MM (male, male) twins were followed by an average of 1·66, MF twins by an average of 2·00 and FF twins by an average of 2·71 further siblings. This result was thought to imply either a reduction in the fertility of women after bearing MM twins or else that MM twins tend to occur later in a woman's life than MF or FF twins. A similar but less clear-cut tendency was found in a study of 249 English families. On the other hand, no such correlation appeared in an analysis of 7420 Mormon families, mainly of the nineteenth century, recorded in the archives of the Mormon Church in Salt Lake City, and a study of records of twinning in the Bola area of New Britain also failed to confirm the results of the Swedish survey. The difference probably gives the clue to the cause of the effect. The modern Swedish and English families concerned may well have been subject to at least some degree of family planning, in which case the occurrence of fewer children after MM twins than after FF twins would be understandable. By contrast, planning is not likely to have played any part in family building in Salt Lake City a century ago or in a remote part of New Guinea recently. The effect is therefore probably a social one, but it is not possible to rule out entirely the interesting biological possibility of an anti-fertility effect of male twins.

In a paper dealing with the birth intervals revealed by the Mormon records, it was recorded that the birth interval between two singletons was influenced only very slightly, if at all, by the sex of the first. Intervals after the birth of twins, however, were substantially longer (Table 4.1) regardless of the sex of the twins or that of the following child. This effect

TABLE 4.1
Birth intervals according to sex and twinning

Combination	Mean interval (months)
M – X	29·28
F – X	29·04
X – M	29·12
X – F	29·21
M – XX	36·24
F – XX	36·51
XX – M	35·38
XX – F	36·54

X denotes child of either sex.

From G. Wyshak. Intervals between births in families containing one set of twins. *J. biosoc. Sci.* **1**, 337 (1969).

may possibly have been a social one, even in the absence of modern methods of birth-control. Wyshak's other findings in this connection, that the interval preceding twins was longer than the interval preceding singletons, irrespective of the sex of the children, is very difficult to explain as a social phenomenon, and is presumably a biological one, possibly depending on the fact that older mothers have more twins but longer birth intervals.

5

Twins, Triplets, and Quadruplets*

Multiple births in man illustrate two interesting biological phenomena—the occurrence of (a) multiple ovulations and (b) polyembryony (embryo-splitting), resulting in multiple pregnancies in an animal which normally produces only one at birth. In this respect man resembles many other animals, notably the other primates and the ungulates, though the frequency of multiple births and of its two different causes varies greatly in different species.

The object here is to discuss the ethnic, geographical, and secular variation in the incidence of multiple births in man and the relative contribution made by multiple ovulation and and polyembryony. I do not propose to consider incomplete splitting of the embryo to produce joined foetuses ('Siamese twins'), in spite of their physiological human interest, or the production of litters by women as the result of maladjusted treatment with pituitary gonadotrophin (fertility drugs).

Frequency of multiple births

It is well known that in most Western countries twins occur once in 80–100 births, so that one person in, say, 45 is a twin. Triplets, quadruplets, and quintuplets are, of course,

* This chapter is based on a lecture given at the Second International Conference on Reproduction in Mammals, held in Nairobi, Kenya, in 1968. In the course of the session devoted to human reproduction, Dr. Ursula Cowgill read a paper on 'The season of birth and its biological implications' from which it became apparent that she took a poor view of children born in September and also of third children. I spoke immediately after her on 'Multiple births in man', and I felt it appropriate to warn the audience not to expect too much because the paper was being given by a third child born in September.

progressively far less frequent and occur, by some happy arithmetical accident, in a fairly regular series of decreasing frequency. Long ago, on the basis of statistics collected in South Germany, it was postulated that twins occurred with a frequency of 1 in 88, triplets 1 in 88^2, and quadruplets one in 88^3 births. Much argument has centred round this alleged law, especially as to the exact ratios involved, but an analysis of huge numbers of multiple births occurring over 10 years in 21 countries gave results surprisingly similar, twins 1 in 85·2, triplets 1 in $87·2^2$, and quadruplets 1 in $87·5^3$ births. Whatever the exact ratios may be, there is evidently a well-defined progression of this kind, which is a very remarkable phenomenon in view of the diverse factors, including pregnancy wastage, involved in determining the number of young produced at a multiple birth. There is no obvious biological explanation, though some have been toyed with. Extrapolation of the series might suggest that one set of quintuplets would appear among about 60 million births.

Embryological composition and sex combinations

Although twins can be of only two kinds (dizygotic (DZ) or dissimilar twins resulting from a double ovulation and monozygotic (MZ) or identical twins derived from the splitting of a single embryo) the embryological origin of higher multiple births can be very much more complicated. Triplets could be derived from three separate fertilized eggs (trizygotic triplets); from two eggs, one embryo having split to form identical twins (dizygotic triplets); or from one egg which has split once to form twins and one-half of which has split again to form a second pair of twins (monozygotic triplets). Splitting of an egg into three to form triplets directly is not apparently known. With quadruplets and quintuplets there are of course many more possible combinations of identical and non-identical individuals; thus quadruplets could be made up of four singles, one single with identical triplets, two pairs of identical twins, or four individuals derived from one egg by a second splitting of each of identical-twin embryos.

Individuals derived from a single fertilized egg of normal

chromosome constitution are necessarily of the same sex and the sex combinations in multiple births are distorted from probability by the occurrence of such individuals. Conversely, the degree of distortion can be used to assess the frequency of occurrence of identical individuals, and for bulk statistics as opposed to obstetrical records this is the only available method. Thus a group of non-identical twins should comprise two males, one male and one female, and two females in the ratio of 1:2:1, apart from any necessary allowance for the slight overall excess of male births. The admixture of 20 per cent of identical twins in the sample, assuming that male and female embryos split with equal frequency, will alter the ratio to three pairs of males, four pairs of mixed sex, and three pairs of females. Considerations of this kind form the basis of Weinberg's rule for calculating the incidence of identical twins, but with the development of sophisticated methods of differentiating DZ and MZ twins it appears that like-sex and unlike-sex DZ twins may not occur in equal numbers, an anomaly which one author[21] has tried to relate to the day of conception in the menstrual cycle.

Turning now to higher multiple births and accepting the usual assumptions, triplets can comprise three males, two males and one female, one male and two females, or three females, and in the absence of identical individuals should occur in the proportion of 1:1:1:1. Admixture of identical individuals of the same sex necessarily confuses the situation. Dizygotic triplets must contain at least two individuals of the same sex, but then so must all triplets, so that the distortion of the expected sex combinations by identical individuals will depend on the ratio of MZ and DZ triplets. With quadruplets and quintuplets the proportion of one-sex births is far above probability, so that a relatively high proportion of the individuals must arise from one egg.

There is no obvious connection between a mother's capacity to produce two eggs at a time and thence DZ twins and the potentiality of an embryo for splitting to produce monozygotic (MZ identical) twins, so that the two types are presumably associated randomly. This assumption is concordant with the fact that ethnic and other variation in the frequency of twinning seems largely to be accounted for by variation in the number of DZ twins.

Ethnic and geographical variation

The figure of 1 in 85·2 (1·17 per cent) for the average frequency of twin births to all births over a period of 10 years in 21 countries has already been mentioned, but the constituent figures varied from about 1 in 53 (1·89 per cent) for Denmark to about 1 in 250 (0·4 per cent) for Colombia. These and other data may be said to indicate a higher rate of twinning in Europe than in America and the Far East, but there are many exceptions. Among countries with a low proportion of twins, Japan was prominent with only 0·69 per cent. Further, the low incidence has been shown to be due to the small number of DZ twins, the proportion of MZ twin births to total births being about the same as in the rest of the world (0·3–0·4 per cent). This very curious fact was confirmed in several later investigations. The fact that Japanese women carry MZ twins as well as women anywhere else indicates that their failure to produce DZ twins in average numbers is not due to pregnancy wastage. We come then to the biologically interesting conclusion that Japanese women have a lesser tendency towards multi-ovulation than have European women, but that their eggs have an equal potentiality for embryonic dichotomy. Elsewhere in the Far East twinning appears to be more frequent than in Japan and within the European range. Thus there are records of 1·1 per cent of twin births in Saigon and 1·08 per cent in Singapore.

Turning to African races, a study of Yoruba mothers in a rural area of Western Nigeria showed that 5·3 per cent of total maternities were twin, and the authors noted that this figure was, in their experience, a world record, especially as there was a high proportion of first maternities in the sample, arising probably from the preferential hospitalization of twin-bearing primiparous women.* The same authors also found that the incidence of twinning rose with the birth order and that there appeared to be seasonal variation; 91 per cent

---

* A primiparous woman is one experiencing her first childbirth.

of the twins were DZ, so that even with the very high overall twin rate the MZ rate was within the normal range. Similar observations have been made in Southern Nigeria, where multiple births were about five times as frequent as in Europe, both DZ and MZ twins having a higher incidence. This state of affairs was ascribed partly to the frequency of higher parities. Such high ratios are not apparently found everywhere in Africa or among Africans in other parts of the world. Among the negro population of Antigua, for instance, both DZ and MZ twinning rates are reported to be within the European range. In the United States, the negroes produce more twins than the white population (Table 5.1), but the rate is still much below those cited above for Africa.

TABLE 5.1

Twinning rates in the United States 1922–36

| Population | Percentage of all maternities | | | MZ as percentage of total twins |
	All twins	DZ twins	MZ twins	
White	1·13	0·74	0·39	34·51
Coloured	1·42	1·01	0·41	28·87

Compiled from figures given by H. H. Strandskov and E. W. Edelen. Monozygotic and dizygotic twin birth frequency in the total 'white' and the 'coloured' U.S. populations. *Genetics, Princeton* **31**, 438 (1946).

In Baltimore, during the years 1941–8, white twin births amounted to 1·01 per cent and non-white to 1·29 per cent of the respective total births. The frequency of MZ births at 0·44 per cent and 0·43 per cent was virtually the same in the two groups, the overall difference being accounted for entirely by the DZ births, 0·57 per cent compared with 0·86 per cent. Adjustment for maternal age and birth order altered these figures slightly, but not in biological significance.

There have been many compilations of figures for Europe. A recent one[4] for 26 countries shows variation in the twinning rate, standardized for maternal age, from 0·91 per cent for Spain to 1·63 per cent for Latvia. For 16 of these countries ranging in total twinning rate from 0·91 per cent for Spain to 1·38 per cent for Greece, the DZ twin rate varied from 0·59 per cent for Spain to 1·09 per cent for Greece, the MZ rate from 0·29 per cent for Greece to 0·38 per cent for Norway. The major influence exerted by the DZ rate is well shown in Fig. 5.1. In this study, most interesting distribution

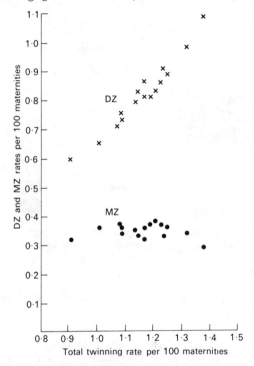

FIG. 5.1.   Total twinning rate and DZ and MZ rates.[4] Variation in the total twinning rate is due almost entirely to changes in the DZ rate.

maps were compiled for twinning rates in Spain, Portugal, and France which, together with the figures quoted above, seem to justify the conclusion of this work, that there is an area of low twinning potential in south-west Europe though even here there is much local variation—for instance, relatively high figures for Spain are found in the sherry-producing district (Fig. 5.2).

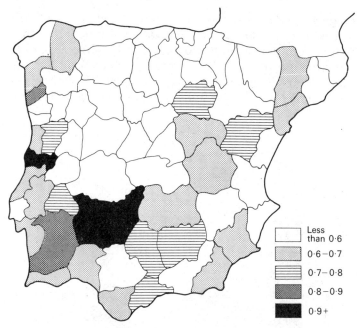

	Less than 0·6
	0·6 – 0·7
	0·7 – 0·8
	0·8 – 0·9
	0·9+

Fɪɢ. 5.2.   DZ twinning rates in Spain and Portugal.[4] Central and northern Spain has a low rate compared with other areas.

Maternal age and birth order

Maternal age is probably the best-authenticated correlate with twinning. The incidence of both single and twin births varies greatly, of course, with the age of the mother, but the peak for twin births occurs some years later than that for single births. Thus, in data published in 1945 for births in a Geneva maternity hospital, 1934–48, the peak for single births was at a maternal age of 25–29 years, that for all twin births at age 30–34 years. These figures conceal a more marked difference in the incidence of single and DZ twins, and a much wider spread of MZ twins for which the peak occurred, as with single births, at the maternal age of 25–29 years. In data from two English hospitals, maternal age groups 25–29 and 30–34 both had about the same incidence of twins; DZ and MZ twins were not distinguished. Of the 475 twin maternities analysed, 175 were first births, which presumably means that first pregnancies diagnosed as twin are hospitalized much more often than later ones.

From a mass of statistics obtained from official sources in Italy, 1949–50, it can be concluded that DZ twinning is related independently to both maternal age and parity (birth order) while MZ twinning is related more closely to maternal age.[28] The total twinning rate was highest in the maternal age group 35–39 and at parities 6 and higher. The crude data plotted in Figs. 1.4 (p. 11) and 5.3. show that the overall DZ twinning rate is more than double the MZ rate and that its variation with both maternal age and parity is far greater. Figures for Australia at three different periods in the last 45 years show a similar maternal age distribution, the peak being at age 35–39, when the twin rates for all maternities at that age group were 1·43, 1·61, and 1·58 per cent respectively for the years 1922–4, 1935–7, and 1947–9. This concentration of twinning in mothers in their late thirties is not apparently peculiar to Caucasians (white races), since it has been reported to occur in poor Chinese women in Singapore, for whom, in the years 1950–3, the overall twinning rate was 1·08 per cent. The rate also increased with parity.

In general, there is a well-established relation between age of mother and parity, and this confuses the analysis of the influence of these two variables on the twinning rate or any other index of human reproduction. The two variables can, however, be separated and in analyses of the Italian records[28] and of material from England, Wales, and New York, it can be concluded that at certain maternal ages the probability that a birth will be multiple tends to increase with the order of the birth. This is not easy to understand, unless it is that the more 'experienced' uterus can carry twins more readily than a less 'experienced' one, so that a twin conception is less often reduced to a single birth by pre-natal mortality.

Secular changes in twinning rates

There is much information about variation in the total twinning rate from one decade or even one century to another, but data relating separately to DZ and MZ twins are much more scanty. In either case, figures for secular change are meaningful only when due allowance is made for coincidental changes in maternal age, family size, and other factors which affect the DZ rate and thus have a major influence on the overall twinning rate, and there may be other influences.

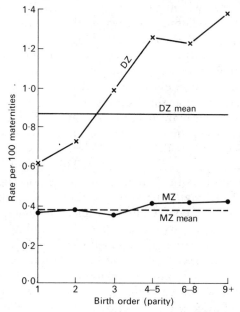

FIG. 5.3.   DZ and MZ twinning rates according to parity.[28] See Fig.
1.4, p. 11 for corresponding effects of age of mother. The
patterns for age of mother and parity are similar, but parity
has some independent influence.

One interesting study showed a high but decreasing rate of twinning in the isolated, Swedish-speaking people of the Aland Islands; over successive 50-year periods the rate varied from 2·23 per cent to 1·52 per cent, the mean over the whole period being 1·91 per cent. The report discussed the decrease in twinning in relation to the decrease in marriages between close relatives, and to the decrease in maternal age and number of children caused by family limitation. Triplets amounted to 0·034 per cent, about twice as many as in the rest of the northern countries.

One of the more interesting investigations in this field[22] dealt with changes in the twinning rate in the United States between 1922 and 1958, for whites and non-whites. From the data the following conclusions may be drawn:

1. Over the period in question there was a substantial overall decrease in white twin maternities from about 1·15 per cent to about 1·0 per cent (Fig. 5.4). However, the decrease was irregular, there being a very sharp drop in the 5 years up to 1943 and an even sharper rise to a peak in 1946. The reason for the drop is not obvious, but if one assumes that food conditions improved materially at the end of the war in 1945, the 1946 peak might be reminiscent of the shepherd's practice of 'flushing' his ewes at tupping time to increase the proportion of twin lambs (there are other possibilities (see p. 111)). By contrast the white

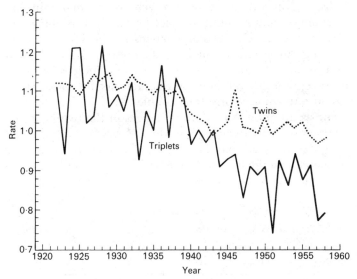

FIG. 5.4.   Twin and triplet rates among the white population of the United States 1923–58.[22] The sharp fall in the twin rate about 1940 was offset after the Second World War by a sharp rise, but the rate then returned to the low level. By contrast, the overall trend in the triplet rates was for a continuous decrease. Twins per 100 and triplets per 1000 maternities.

triplet rate declined more steeply and, on the whole, more
regularly than the white twin rate.

2. For non-whites the twin picture was rather different. In
keeping with the data given earlier the overall rate was
higher than for whites. The overall decrease over the
period considered, from 1·47 per cent to 1·37 per cent,
was less for non-whites than for whites, but the decrease
in the years before 1943 was even more marked and the
subsequent recovery slower and more sustained.

In view of the relative stability of the MZ rate compared
with that of the DZ rate, it would be reasonable to assume
that the secular changes appearing in the United States total
twinning rate were due to variation in the DZ rate, and possibly to changes in maternal age and family size. This explanation was accepted in one review of the incidence of multiple
births, but the authors of the American report maintain that
the changes cannot in this instance be ascribed to changes in
these variables. The secular changes in the United States
twinning rate are thus something of a mystery. It may be that
obscure social factors are involved, such as increasing use of
contraception differentially by the more fertile and perhaps
therefore the more twin-prone couples.

The assumption that secular changes in the twinning rate
are due primarily to changes in the DZ rate is not borne out
by figures for Australia (Table 5.2), which seem to show that
the increase in the twinning rate between 1922 and 1949 was
due very largely to increase in the MZ rate, but the change

TABLE 5.2

Total twin births and estimated MZ and DZ twin births, Australia

Period	Total twin births	MZ births	DZ births
1922–4	1·032	0·308	0·724
1935–7	1·065	0·349	0·716
1947–9	1·111	0·374	0·737

Rates per 100 maternities. After McArthur. Changes in twin frequencies.
Acta Genet. med. Gemell. **3**, 16 (1954).

was very slight compared with those caused elsewhere by variation in the DZ rate. The figures for the MZ rate, as in so much of this work, were calculated from the excess of like-sexed twins over what would be expected for DZ twins.

A further side-light on the results of the American investigation is thrown by analyses showing that the DZ twinning rate fell during the Second World War in France, Holland, and Norway, but not in Denmark, Sweden, or north-west France. The MZ rate remained constant in all these countries. It is suggested that, where the DZ rate fell, the tendency to dual ovulation was undermined or the capacity of the organism to implant two blastocysts impaired by malnutrition. Factors affecting an established pregnancy would presumably affect DZ and MZ twin pregnancies equally.

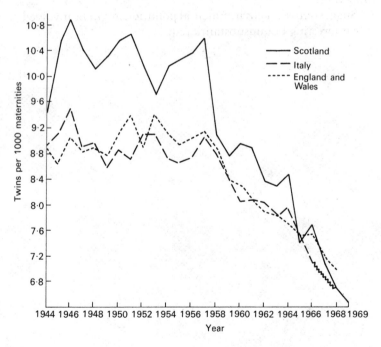

Fig. 5.5.   DZ twinning rates in Scotland, England, and Wales, and Italy 1944–69.[21] The fall started some 20 years after that in the American rate.

All such changes, however, were comparatively slight compared with the decline in the DZ rate since 1958 in many European countries[21] which, in England and Wales for instance, brought the rate down from around 0·9 per cent for the previous decade to about 0·68 per cent by 1968–9. In Scotland, the fall reached about the same low point, but from a much higher level (Fig. 5.5). Outside Europe, both Australia and New Zealand shared in the decline. The United States, having experienced a major fall 20 years earlier, was little affected by the changes of the 1960s. Earlier marriages and younger mothers with fewer twins must have played a large part in the later fall in the DZ twinning rate. Whether this is the whole explanation, that is, whether the age-specific rates failed to show any significant change, is uncertain.

Suggestions that environmental pollution might be involved are interesting but unsubstantiated.

6

The sex ratio

We are all so conditioned to the occurrence of the two sexes in approximately equal numbers that we rarely ask how this comes about, why the ratio bears so little relation to the relative reproductive capacities of the two sexes, what influence it has on our lives, and what is the extent and the cause of the variations which occur. The first three questions are all interconnected, and in this context can be answered briefly. The last one calls for more discussion because of the changes in the sex ratio during the life-cycle and the large number of factors which influence the sex ratio at different times and in different places.

The chromosome mechanism

The sex of a mammal is determined by its chromosome constitution, which is $2\mathcal{N}+XX$ in the female and $2\mathcal{N}+XY$ in the male, where $\mathcal{N}$ is the number of paired chromosomes, X a sex chromosome, and Y a small or rudimentary sex chromosome. Among the cell divisions which occur when the germ cells are formed one separates the chromosomes of each pair, giving a spermatozoon or ovum with half the full number. In the female, therefore, all the ova are of the same type $\mathcal{N}+X$, while in the male, spermatozoa are of two types, $\mathcal{N}+X$ and $\mathcal{N}+Y$. On fertilization, therefore, not only is the original number of chromosomes restored, instead of being doubled, but the sexes are reconstituted, on a chance basis, in equal numbers thus:

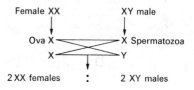

Mechanisms of this kind exist at most levels of the animal kingdom, but in many the results are subject to modification by a variety of factors. In mammals, including man, modification of the normal results of the sex-determining mechanism is comparatively infrequent. So far as the gametes are concerned, therefore, the woman is not responsible for the sex of her child.

In animals in which the two sexes play a more-or-less equal part in reproduction, a sex ratio of 1:1, as provided for by the chromosome mechanism, is perhaps understandable. By contrast, where the development of maternal care by gestation and lactation has severely limited the reproductive potential of the female without affecting that of the male, a 1:1 ratio does not provide maximum reproductivity per head of population. Maximum reproductivity is presumably a survival factor, and one may well wonder therefore whether the 1:1 ratio has been subjected to pressures during the evolution of the mammals, and if so whether the surplus of males has persisted because it increases genetic variability and thus promotes natural selection, or simply because of the difficulty of altering the sex-determining mechanisms. In either case, the human race is stuck with the mechanism, and thence with the 1:1 ratio, until such time as it can be altered artificially. In the meantime our social system will continue to be profoundly influenced by the approximate numerical equality of men and women.

Little thought is needed to visualize the sort of society which might have evolved around a sex ratio more in line with the reproductive capacity of the two sexes, say one male to ten females. A truly polygynous society of this kind does not exist in present circumstances because even where polygyny is still practised the males left over from the 1:1 ratio have either to be disposed of or left as a disturbing bachelor element in the social hierarchy. I have suggested elsewhere other social consequences of a more ergonomic sex ratio.

It follows from the chromosome mechanism that the sex of the new individual is determined at fertilization and, barring rare developmental abnormalities or external interference this cannot be changed. Even chromosome abnormalities caused by non-disjunction of the sex chromosomes or otherwise, producing for instance XXX or XXY individuals, usually imprint a genetic sex on the individual. There is thus at fertilization a sex ratio which can be changed only by sex-reversal or by differential mortality. Complete reversal of genetic sex in mammals must be very rare, even if it occurs; differential mortality, on the other hand, exerts a profound effect on the sex ratio throughout the life-cycle. In mammals, and other groups of animals, the male appears to be less viable than the female, and to a greater or lesser extent to have a higher mortality rate at all stages. As a result, the sex ratio, expressed as the number of males per 100 females or as the percentage of males, declines during the life-cycle. In man, this differential mortality of males was very evident during infancy in the days when infant mortality was high and it is still seen both then and later. It is also probable that there is a differential wastage of males in pre-natal life, and that the ratio at conception is higher than that at birth.

The sex ratio at conception and during pre-natal life

The chromosome mechanism implies that X spermatozoa (female-determining) and Y spermatozoa (male-determining) are produced in equal numbers, but the ratio could easily be upset by differential wastage at any later stage, especially during maturation and storage of the spermatozoa in the male tract. Whether or not this happens is unknown because there is no way of identifying and counting the two types. We do know, however, that there is a heavy overall wastage of spermatozoa in the male tract, which could well be differential. It is only an assumption, therefore, that the X and Y spermatozoa leave the male in equal numbers. It is even more doubtful whether they reach the site of fertilization in the female tract in equal numbers. What we do know is that of the scores of millions which are deposited in the vagina at coitus only a few hundreds or thousands have the luck and vigour to make the long journey to the Fallopian tube. There is here unlimited opportunity for differential elimination, and it has been suggested many times that the Y spermatozoa may have a slight advantage of motility or survival (see also p. 115). All this adds up to the fact that we simply do not know what sort of sex ratio to expect among the new individuals at fertilization. Opportunities for estimation of the ratio at this stage are virtually non-existent in man, and estimates in experimental animals can be made only with difficulty.

The segmenting egg or early embryo can be sexed, according to present knowledge, only by observation of the sex chromatin, the small knob protruding from the nucleus of the cells of the female, or by preparing nuclear squashes for examination of the sex chromosomes. The sex-chromatin technique suffers from the difficulty of making a clear-cut identification in every case, and the chromosome technique requires pre-treatment of the tissues with colchicine to arrest cell division at the right stage. For these and other reasons the use of either technique in the very early stages of development presents many difficulties.

Progress in sexing tubal eggs or embryos soon after implantation is obviously going to be slow; in the meantime we still have the time-honoured method of determining the sex ratio by microscopic or macroscopic examination after overt differentiation has taken place. This stage, of course, is reached comparatively late in pregnancy in most species, and, until the development of the sex-chromatin and chromosome techniques, our knowledge of the pre-natal sex ratio related only to such late stages.

This brings us to the crux of the pre-natal problem. To estimate the sex ratio at different stages of pregnancy we can either sample the intra-uterine population directly, or we can take the ratio at birth and calculate backwards according to the amount and sex incidence of the mortality known to have occurred at any particular stage. The first of these methods was used in early work on pigs and cows. The results clearly indicated a higher ratio during pregnancy than at birth, and, in the case of the pig, an inverse relationship with the size (that is, the age) of the foetuses, the older the foetuses the nearer the ratio to 1:1.

Sampling of the intra-uterine population

In man, until recently, the only available sample of the intra-uterine population was that provided by spontaneous abortions. Material of this kind has usually been reported as showing an excess of males, but such a sample is not likely to be representative of the uterine content, though it may be useful for indirect estimates of the sex ratio at conception. In any case, spontaneous abortions may include a lot of abnormal material. Analysis of the chromosome constitution of a series of spontaneous abortions revealed 95 normal males, 104 normal females, 11 mongoloid males, 23 mongoloid females, and 22 cases of Turner's syndrome. The high proportion of abnormal foetuses, about 20 per cent, is remarkable, as is the overt sex ratio; a mixed bag of this kind, sexed by macroscopic inspection at a later stage, would have yielded 106 males and 149 females.

The direct sampling of the intra-uterine population in man has become possible only since the introduction of mass induced-abortion programmes. Another investigation, based on chromosome examination of induced abortions, gave 102 males and 92 females, a ratio of 111:100, presumably in the first third of pregnancy. This figure is nicely confirmatory of previous estimates. Data for Czechoslovakia, where abortion for social reasons has been practised for many years, suggest a large excess of males in the early stages of pregnancy (Table 6.1).

Indirect estimation of the pre-natal sex ratio

The indirect method of working from the sex ratio at birth and the amount and incidence of pre-natal loss has naturally been much used for the human subject. Here, naturally, we start with the fact that there is a slight excess of males at birth. We must then ask whether this derives (a) from a 1:1 ratio at conception, subsequently modified by differential female mortality, (b) from a raised ratio at conception unaltered by equal loss of the two sexes during pregnancy, or (c) from a high ratio at conception, reduced by differential mortality of males during pregnancy. Almost all the earlier work on humans pointed to a high ratio at conception because of the apparent high sex ratio of abortuses and the known higher ratio among stillbirths, both being concordant with the lower survival rate of the male in post-natal life. The arithmetic of such calculations is at least plausible. If we take a ratio at birth of 105:100, a ratio of 150:100 among abortuses and a wastage of 25 per cent during pregnancy, we get the approximations shown in Table 6.2, which suggests that there might be 115 males : 100 females at conception.

TABLE 6.1

Sex of early human foetuses

Foetal month	Sex of foetus		Total	Males per 100 females
	Male	Female		
2	57	32	89	177
3	253	185	438	136
4	68	58	126	117
Total	378	275	653	137

From O. Hnerkovaky, E. Petrikov, and M. Cerny. Pre-natal sex ratio in men. *Acta Univ. Carol. Med.* **18** (Suppl.), 109 (1964).

It is evident, however, that the sex ratio or the amount of wastage would have to be appreciably higher to push the ratio at conception much above 120.

TABLE 6.2

Sex ratio at conception

	Males	Females	Total
At birth	105	100	205
Wastage ($\frac{1}{3}$ of 205 and $\frac{3}{5}$ males)	41	27	68
At conception	146	127	273

Investigations during the 1950s largely substantiated such calculations, as shown by the following summary of the conclusions of several authors, some using sex-chromatin sexing:

(a) the sex ratio of all abortuses is high, about 135♂:100♀ according to one author;

(b) the abortion rate is highest between the second and fourth months of pregnancy, and the excess of male abortuses greatest at this stage;

(c) estimates of males per 100 females at conception vary between 124 and 107.

The sex ratio at birth

Since the sex ratio at birth will be the result of the ratio at conception modified by the amount of sex incidence of pre-natal mortality, variation in the ratio will depend on variation in these two components, which in itself will depend on largely unknown factors.

There is another point. Several of the variables which apparently affect the sex ratio at birth are interrelated. Thus if younger mothers produce relatively more males, then anything that promotes earlier marriage will also apparently result in more males being born. Moreover, as earlier births come more from younger mothers than later births, there will apparently be a relation between birth order and sex ratio. It follows that the two component factors which determine the sex ratio, and the probable interrelation between variables which apparently affect it must be borne in mind when considering deviations in the sex ratio at birth.

Ethnic variations

Information under this heading is not as abundant or as reliable as could be wished, because in many parts of the world where deviations might perhaps be likely, birth records are not available or are biased by the preferential registration of male births. It seems safe to say, however, that no startling departures from 1 :1, say less than 90:100 or more than 110: 100, have been recorded on the basis of adequate figures for whole populations. A further generalization is that in most areas there is a greater or lesser excess of males at birth, which varies from time to time and place to place. The sex ratio at birth, therefore, varies not around numerical equality, but around a significant degree of masculinity.

Records of 46 million live births in western Europe from 1929 to 1937 showed 51·3 per cent of males, or 105·3 males per 100 females, and the author went on to say that 'data from other continents are less reliable but suggest a high degree of uniformity throughout the human race'. Later, 51·4 per cent of males was found in 21 million births among whites in the United States during 1946–52. In the Far East, samples of births in Korea and Japan yielded male percentages of 53·5 (115:100) and 52·2 (109:100) respectively, and smaller samples in Samoa and New Guinea had percentages of 51·3 and 51·9. In the Argentine more than a million births in Buenos Aires between 1929 and 1943 had a ratio of 105·4 males for 100 females; extremes were 103·6 in 1938 and 107·3 in 1939.

The best authenticated records of apparent ethnic difference in the sex ratio at birth come from the white and coloured populations of the United States. All of the many observers have reported a lower proportion of male births among negroes. One of the later reports gives a male percentage of 50·7 among about 5 million negro births, compared with the 51·4 recorded above for whites. Similar figures have been given for other parts of the world where large numbers of white and coloured people are living under similar climatic conditions, for example, Cuba and South Africa. To what extent the differences are caused by social conditions rather than ethnic characteristics is uncertain, but social conditions might well affect pre-natal wastage, and thence the proportion of males at birth. Most observers agree that geographical and climatic factors as such have little effect on the sex ratio, but there have been contrary reports. For instance, it has been suggested that a slight masculinity gradient exists west and north across the Pacific. A more local and extreme report relates to Europeans in Nigeria, who have been reported (in 1957) to produce a normal proportion of males in the healthy highlands, but a low one in the hot, humid, tropical parts of the country.

Secular variation

Year-to-year changes in the sex ratio are well known. For instance, the ratio in Finland is said to have risen continuously from 1751 to 1948. The extent of secular changes should not, however, be exaggerated. A table drawn up in 1884 showing the sex ratio at birth in 32 states or countries listed only two outside the 105–6 range, ratios very similar to those found today.

In England and Wales the male excess has increased irregularly during the present century from rather below 104:100 to around 106:100, with two sudden increases following the two World Wars. This effect of war is well known and occurred elsewhere in Europe at the same time; a ratio of 112 males per 100 females was recorded in Düsseldorf in 1946–7. Such a rise has also been noticed with lesser wars, and for a long time it was thought to be a heaven-sent device for replacing the males lost by war. The more modern and more prosaic explanation is that it results from the earlier marriages and younger child-bearing which follows wars. In the United States, the sex ratio at birth was said in 1938 to have fallen slightly over the previous 30 years, but as elsewhere it was raised at the end of the Second World War. There is evidence, however, that in this case the rise was not associated with earlier marriages (see also p. 112).

Seasonal variation in the sex ratio has also been reported for the United States, where it seems that the highest proportion of males occurs in the period May–July and the lowest during October–March (see Fig. 8.1). In 1952, the seasonal fluctuation in the sex ratio of births in Japan was almost the reverse of the American pattern, the lowest proportion of males being found in December–April and the highest in October–November. In March there was an actual excess of females, while in November the male excess reached 112·5 : 100. As an extreme of secular variation it has been recorded that more males are born before sunrise and sunset, and more females around midday and midnight.

Parental age and number of previous children

There is obviously a close connection between parental, especially maternal, age and the number of children in a family (the parity), and their relative contribution to changes in the sex ratio can be estimated by considering the same parity at different ages, and vice versa. The general conclusion is that the proportion of males born decreases both with increasing age of mother and increasing parity, and most authors have regarded both of the variables as important. According to some, however, when parity and age of mother are each stratified in considering the effect of the other it becomes clear that parity is the operative variable, and the apparent effect of age of mother is incidental to this.

It has been maintained, in fact, that the age of the father is a more potent factor than that of the mother. Whatever the precise correlates, however, there is no doubt about the high sex ratio of births to young mothers and primiparae (first-time mothers), ratios of around 115 : 100 females being frequently recorded. Nor is there any doubt about the low sex ratio among births to old multiparae (mothers with many children). Whether or not this decrease in the proportion of males can be attributed entirely to the higher pre-natal loss which occurs under the same conditions is uncertain (see p. 92). Information for stillbirths is, naturally, much more available; the stillbirth-rate doubles with high maternal ages and the proportion of males among stillbirths also increases.

There is one rather curious report in this connection.

Among Japanese women, it was noted in 1952 that up to a maternal age of 50 the changes in the sex ratio at birth were broadly similar to those described for other countries, but that in maternal groups 50–54 and 55+, the sex ratio of births rose sharply to 111 and 119 respectively. For the sake of the mothers and children one may hope this phenomenon is based on inadequate material or on misregistration (see p. 21).

Sex combination in families

Assuming an overall ratio of, say, 105:100 it is possible to calculate the probable frequency of the various possible combinations of the two sexes in families of any given size. Many investigations of this kind have been made, giving rise to the general conclusion that families consisting of one sex only, or largely of one sex, are more frequent than would be expected from probability. This is not easy to understand. The complementary fact that the tendency is less marked in small families and more in large ones is evidently because most people wish for a family with at least one of each sex and tend to continue production until this is achieved. In some instances of course, the desire is not so much to produce one of either sex, but to produce an heir, in which case small families should have an excess of males and large ones an excess of females.

Sex ratio in multiple births

The sex ratio among multiple births in man is significantly disturbed by the occurrence of identical twins which are necessarily of the same sex. The resulting complications are discussed on p. 75.

Other factors

Many other factors have been said to affect the sex ratio at birth. The effect of race-crossing, for instance, was the subject of considerable study. Results, however, are in doubt, and it should be emphasized that race-crossing, like other factors, must operate through the ratio at conception or through differential mortality afterwards. Evidence for alteration of either factor due to biological effects of miscegenation is scanty. Social effects on pre-natal mortality, of course, may well be an operative factor.

Other variables credited with influencing the sex ratio at birth include inbreeding, polygamy, nutrition, illegitimacy, and relative age or vigour of father and mother. At one time, a good deal of attention was given to a supposed relation between illegitimacy and sex ratio, the idea being that attempts at concealment and so on would increase pre-natal mortality and thence decrease the sex ratio at birth. Whatever may be the case during the nineteenth century, the Registrar General's figures for modern times give no suggestion of anything abnormal about the sex ratio of live-born illegitimate children. All that need be said here is to emphasize that such factors, like the more important ones, if they operate at all, must do so through one or both of the two determining factors.

The sex ratio from birth to death

We have seen that a slight excess of males at birth is the general rule for the human race. In Great Britain the ratio at birth is now about 106:100. In the past, the excess of males at birth was short-lived, being eradicated, mostly before the end of the first year of life, by the differential incidence of the high infant mortality. Thereafter male mortality continued to exceed female mortality at all ages except in the 10–14 age group, at which age a slight excess of females died, possibly because of the more severe stresses of puberty. The end result was a considerable excess of females at marriageable age. In those days, therefore, the male had a slight scarcity value. At the present time the situation is very different. Not only are slightly more males born than in the past, but because of the greatly decreased mortality in early life, an excess of males is preserved into the 25–29 age group, and approximate equality is reached only in the 30–34 group. Thereafter there is an excess of females which increases steadily throught the rest of life, so that women outnumber men by more than 2:1 among old people of 80 or more. Up to the age of 30 or so, however, females now have a slight scarcity value. Let us hope that they make better use of it than did the males in their heyday. One result is already evident. Twenty years ago when the trend towards a shift in the sex ratio of young adults was already evident, I prophesied that the males would have to do something to make themselves conspicuous. It is nice to be a true prophet!

This process of preserving the excess of males born will no doubt continue and may well be reinforced by a decrease in the pre-natal wastage of males. We should look forward, therefore, to a future in which a substantial excess of males will persist right through the reproductive phase of the life-cycle. With similar trends showing in other countries, we are heading for a world shortage of marriageable females.

7

The social impact of human reproduction

Man is essentially a gregarious animal and is subject, therefore, to both biological and social pressures. As a result,
almost everything he does has biosocial repercussions. This is
especially true of reproduction, which is a biological process
with an overriding social impact. If human reproduction
came to an end, so would the human race. By contrast the
present unprecedented combination of a high reproductive
rate and a high survival rate is causing widespread concern.
One may take the view that mankind, by reason of its explod-
ing numbers, is rushing to a Gadarene doom, or the opposite
view that man has brains as well as gonads and will cope with
his proliferation. In either case, one must admit that at the
present time there is a population problem.

The reason for the problem is obvious. In nature every
species combats a low survival rate with a high reproductive
rate. In man, in modern times, medical science has raised the
one without, so far and taking the world as a whole, having
done much to lower the other. The result, as we all know, is
an increase in world population which simple arithmetic
shows cannot be sustained indefinitely. This conclusion was
authoritatively pinpointed in a forceful report by a Committee
of the National Academy of Sciences in Washington published more than 10 years ago. The verdict of the Committee
was that 'either the birth rate must come down or the death
rate must go back up'. In this situation, which in essence is
quite simple, we have our greatest biosocial problem; it is
essentially one of adjusting reproductive potential to a level
of reproductive performance appropriate to modern conditions of survival.

Man's reproductive capacity

First of all, what is this reproductive capacity which, in association with a high survival-rate, is so alarming? Given a normal sex ratio, the average production of one surviving female child per woman could result in a potentially stable population. By contrast, the average production of two surviving female children could double the population every generation, say every 25 years. A geometric increase of this kind would obviously be catastrophic in an historically short time. Yet human reproductive capacity is well beyond the comparatively modest output of two female children per woman.

A fertile male, during his lifetime, produces untold billions of spermatozoa, and under conditions of polygyny, because of his minor contribution to the reproductive process, could father, and some undoubtedly have fathered, many hundreds of children. Similarly, female reproductivity is not usually limited by the supply of germ cells (see p. 8).

What then restricts the reproductive capacity of the human female? Obviously her capacity for successive gestations and for carrying multiple pregnancies. The world record is said to be held by a Russian peasant woman who is reputed to have produced, in the middle of the last century, 69 children in 27 confinements: 16 pairs of twins; 7 sets of triplets; and 4 sets of quadruplets. This prolific woman was, moreover, only the first wife of the children's father. We may well regard this story as on a par with that of Russian sesquicentenarians fathering children, but Great Britain is well up with the field. The current record appears to be a mere 22 or so, but according to the *Guinness Book of Records* a woman living in Hertfordshire in the seventeenth century married at the age of 16 and had a world record of 38 confinements in which she produced 32 daughters and 7 sons. Her last son, who died around 1740, became a distinguished surgeon and wrote a book on embalming. On a recent visit to Peru I saw a newspaper headline which, translated, said 'Mother of 24 children refuses the Pill'. The Hutterite communities in the United States have also put up some remarkable records.

Man's reproductive capacity, therefore, though small in relation to his supply of gametes, is still very great. Why then, has the human race not exploded earlier in its history? That answer is well known.

In nature every species, however high its reproductive rate, is held in check by limiting factors of which the main ones are food supply, disease, and predators, and there can be no doubt that, until comparatively recent times, such factors, aided often by restrictive social customs, were extemely effective in checking population growth.

It took up to half a million years (depending on its antiquity) for the human race to reach an estimated 300 million in the year A.D. 1000 and more than another 800 years to reach an estimated 1000 million early in the nineteenth century. Thanks to the efforts of medical and agricultural scientists we are now rushing up to an estimated 7000 million by the end of the twentieth century. Such an historically sudden increase in the numbers of so large an animal as man, whose biomass of perhaps 250 million tons is by far the largest among living animals, is indeed a remarkable biological phenomenon and would have been impossible had man remained a food-gatherer and hunter like other animals. Remarkable or not, the increase cannot go on indefinitely. Something will happen to stop it. The question is: what will that something be? Is human fertility declining spontaneously or will it do so under the unnatural pressures of civilization and overcrowding? Will it be left to the limiting factors of nature to assert themselves more vigorously than at present? Finally, and this is the supreme question: will man learn consciously to adjust his fertility to the potentialities of his environment?

Is human reproductive capacity declining?

So few communities now reproduce to their biological limit that it is difficult to answer this question comprehensively, but communities such as the Hutterites offer no suggestion of any decline in human fertility. For the individual, potential fertility is certainly not declining to a level appropriate to the present and prospective survival rate. In some ways it may even be increasing. Puberty is arriving earlier than it used to do (see p. 18), and in our permissive society the fertility of mere children is being amply demonstrated.

In the United Kingdom recently a girl of 12 years old had a therapeutic abortion; the boy responsible was said to be a year older, and in 1971 about 4000 girls under the age of 16 years were known to have become pregnant and no doubt many more were unrecorded. Even 20 years ago, well before the start of the present free-for-all, a few dozen fathers under the age of 15 years were recorded each year in the United States, corresponding to fertile intercourse around the age of 14 years or earlier. The ultimate in fertility is held by a Chilean girl 8 years of age who was delivered of a live child by Caesarian section and so became the youngest recorded mother in human history. This girl started to menstruate at 5 months old and was almost certainly a case of precocious puberty. At the other end of the scale, there is no satisfactory evidence that the age of menopause is increasing (see p. 20) but, even so, a woman in the United States a few years ago, is said to have produced a daughter at the age of 57.

In a different way, the time taken to conceive when conception is desired should be a good index of fertility, since it involves the reproductive efficiency of both male and female. It is difficult to get firm data, but one author concluded that 3 out of 4 couples not using contraception effected conception within 6 months, the median time being $2\frac{1}{2}$ months. Figures of 30 per cent within a month and 90 per cent within a year have also been recorded. In a recent pilot survey, 4 of 11 girls studied in detail when requesting a legal abortion had become pregnant from a single or isolated coition. Other figures have, of course, been very different. One estimate was that more than 259 acts of uncontracepted intercourse were, on the average of all ages, required to induce a pregnancy; even with young couples, supposedly at the most fertile age, the number was reduced only to 175. This may be correct, but the figure is likely to raise a smile on the face of any woman who became pregnant on her honeymoon. All in all, therefore, it would be difficult to maintain that the fertility of the human couple shows signs of declining.

Demographically, of course, there is plenty of evidence of a decline in human fertility in the sense of reproductive performance. The decline in the birth-rate and family size in Europe in the last 100 years is sometimes cited as evidence of declining human fertility; it is certainly an example of declining demographic fertility, but does this have any connection with biological capacity?

Apart from such factors as late marriage, a decline in human reproductive performance can be due only to a decline in fertility, including a decline in sex drive, to negative or positive contraception, or to induced abortion. Few, I think, would suggest that an epidemic of abstinence, a catastrophic decline in sex drive, or a failure of other components of biological fertility started in Europe in the nineteenth century. Negative contraception, in the form of *coitus interruptus*, was no doubt prevalent, and it still is. Positive contraception was primitive and must have been highly unreliable—one recommended recipe was a vaginal sponge soaked in brandy. There was also, it seems, a brisk trade in appliances made from animal tissues, the aesthetic impact of which can be judged from the fact that they were commonly referred to as 'machines'. Positive birth-control, as opposed to positive contraception, in the form of induced abortion, must also have been resorted to frequently, and no doubt infertility arising from do-it-yourself methods added its quota to the decline in family size. All in all, therefore, it is not difficult to ascribe the fall in the birth-rate in Europe to causes other than a decline in biological potential.

Similar considerations apply to the decline in the birth-rate in England and Wales during the first 30 years of the present century, when family size decreased from around 4 to 2—an all-time low. A Committee of the Royal Commission on Population (set up in 1944), of which I had the interesting experience of being a member, could find no evidence that the decline was biological in origin, so that it was presumably due to voluntary birth-control by some means or other. This conclusion was confirmed by the fact that the Committee's deliberations were cut short by the post-war baby boom.

If, therefore, the potential reproductivity of mankind is not decreasing spontaneously, is it likely to do so in future in response to evolutionary, physiological, or psychological factors inherent in civilization? We have, for instance, the intensive campaign to damp down the fertility of fertile people and the almost equally intensive campaign to remedy the infertility of infertile people. Will these paradoxical efforts cause high fertility to lose its survival value, and, if so, will the ultimate result be a decrease in biological fertility? The answer will obviously depend on many factors, including the extent and nature of infertility, especially of its genetic components, and the extent of the resources which, under present conditions of population pressure, society is willing to devote to remedying it.

Cost–benefit studies are often used to reinforce the massive case for the limitation of fertility, and similar studies will some day have to be used to rationalize the treatment of infertility. For instance, it is difficult to imagine that, for the community as opposed to the individual, procedures such as egg transfer and *in vitro* fertilization, however interesting biologically, could be justified on a cost–benefit basis. Much the same applies to the induction of ovulation by treatment with gonadotrophins, for which elaborate control facilities are necessary to avoid super-ovulation and the traumatic experience for the woman of producing a litter. So far, the record in this direction appears to be held by an Italian woman, who is reported to have had 15 foetuses removed at the fourth month following hormone treatment.

One thing, however, is clear. Whatever may be the evolutionary result of present trends, not even the most optimistic observer can think that human fertility will decline, in response to a loss of survival value, in time to solve the present population problem. The re-direction of human evolution, to a point at which the average woman could produce only two or three children in a lifetime, is not likely to happen quickly, if ever. But it is possible that the slow processes of evolution will be speeded up by changes in the physical and social environment. We hear a lot about pollution, especially about the chemical insults to which we are constantly subjected. Will these affect human reproductive capacity and, if so, will they do so in time to supplement conscious efforts towards limiting human fertility? It is impossible to say, but there are suggestive reports, as of impotence among pesticide workers.

Of possible relevance here is the decrease in the DZ twinning rates which has occurred in many countries, especially in Europe, since 1957 (see p. 85). Couples get married earlier than they used to do, and younger mothers have fewer twins than older ones, but this seems not to account for the decrease everywhere. We have here, therefore, evidence of a decline in the frequency of twin ovulations, or in the frequency of coitus necessary to promote the fertilization of two eggs, or in the capacity to carry a twin pregnancy. In short, there may here be some evidence of a decline in fertility. We have no clue as to the cause, but in this connection it is difficult to avoid thinking of the increasing use over the decade of insecticides, herbicides, and growth stimulants.

The conscious control of human fertility

Endogenous adjustment of potential human fertility to meet modern conditions of survival cannot, therefore, be expected for a very long time, if ever. What then of the future? Few people would wish to see the problem solved by the harsh limiting factors of nature or by primitive forms of social control such as human sacrifice, infanticide, cannibalism, self-immolation and the casting out of the old and sick. The inescapable conclusion is that man must learn, and learn faster than he is doing at present, consciously to control his own fertility.

How is this to be done? Let us consider first the oldest method. Abortion for social reasons is now legalized in many countries, and one must applaud this triumph of humanity and common sense—provided that the method is used appropriately. In my view, abortion should be reserved for cases of unforeseen medical or genetic emergency, as a last resort when contraception has failed or for a crash programme of population limitation when the slow spread of contraception would be inadequate. By contrast, I should be very sorry indeed to see so biologically wasteful a process become a routine method of fertility control or to see the use of contraception become more haphazard because of the availability of abortion as a longstop. In saying this, however, I am thinking of indiscriminate abortion without regard for the sex or condition of the foetus, and possibilities are looming up which may make selective abortion socially valuable and scientifically important. I refer to the possibility of diagnosing both the sex and the chromosomal constitution of the embryo by examination of amniotic cells.

Diagnosis of sex at an early stage would permit the removal of an embryo of the unwanted sex, so that people sufficiently motivated to sacrifice an embryo on the altar of the sex ratio could literally choose the sex of their children. With improving techniques this method might be easier and more certain than the alternative possibility of *in vitro* separation of X and Y spermatozoa combined with artificial insemination. Any practicable method of choosing the sex of children would initiate a most interesting social experiment. It is probably correct that the first result would be a small reduction in the birth-rate, because at least some children are produced in the hope of adjusting the sex ratio of the family. By contrast, however, I think the net result on the sex ratio at birth would be slight, because the present ratio of near 50:50 is probably what most parents in the long run would choose for their families. A fashion for boys, for instance, would almost certainly be followed by a fashion for girls, who would have acquired a scarcity value during the glut of boys, and vice versa. I think there would be little chance of the sex ratio adjusting to the reproductive potentials of the two sexes, say 1 male to 20 females, and little chance therefore, of the social and demographic upheavals which the appearance of such a ratio would cause. But if the swings of the fashion pendulum were slow, the social results could be fascinating to watch. A mild experiment of this kind has, in fact, been carried out in front of our eyes in the last 25 years, during which the sex ratio of young adults has swung from an excess of females to an excess of males (see p. 98).

Turning to more serious matters, we have the other possible use of selective abortion—the elimination of at least some types of defective embryos. Diagnosis by amniocentesis would, of course, have its limitations, but it could nevertheless be extremely valuable in reducing the number of defective children. Today, the number of such children is relatively large, and their catastrophic impact, especially on the mother, is exacerbated by the misguided efforts of the medical profession to preserve anything and everything that is born alive, at whatever cost in distress and frustration to the family and in cash to the community.

I shall add very few words to the millions which are written every year about contraception. It is sometimes said that the development of the perfect contraceptive method would solve the population problem. This I do not believe. For one thing, there never will be a method perfect for use at all times, under all conditions, in all parts of the world, and in all cultures. For another, the best of methods will be of no value if it is not used, and here we have to face the problem of motivation.

I have already referred to the alarmingly low birth-rate in the United Kingdom during the 1930s. By that time, knowledge of contraception was spreading widely, but available methods were relatively crude and aesthetically objectionable to many people. The conclusion to be drawn from our low birth-rate, therefore, is that if people want small families they will have them, however primitive the methods of control available; conversely, one may assume that large families will not entirely disappear whatever disincentives are applied to, or contraceptives dangled in front of, the parents. Parental and especially maternal urge is strong; even now there are wives who are not happy unless they have young babies to look after. And there will always be 'accidents' even after the desired family size has been achieved. The capacity for forward planning is not universal; few people have a brain packed in ice during the count-down to coitus, and we hear a lot these days about the emotional stimulus of risk-taking, erstwhile known as 'taking a chance'. Moreover, in some ways man is still a primitive creature, as witness the peak conception-rates in the United Kingdom and the United States around the summer and winter solstices respectively. In England, the peak occurs during the summer holiday period, in spite of the fact that there is a coincident peak in the sale of contraceptives.[86]

Motivation, therefore, not contraceptive technology, is the main bottleneck in fertility control, and the situation is complicated by the fact that motivation now has two distinct aspects—the welfare of the family and the welfare of the human race. Family planning was initiated with the idea of limiting and spacing births in the interests of the mother and the family. It had no concern with national or global demographic problems, and even good family planning can result in a size of family much in excess of that implied by considerations of population growth. Ought couples, therefore, to consider wider issues than their own circumstances in deciding how many children they should have? This is not a new question. A. V. Hill in his Presidential Address to the British Association for the Advancement of Science in 1952 specifically asked whether, under modern conditions, couples had the right to unlimited reproduction. In the years since A. V. Hill's address, the far-sighted question has often been debated, but no clear answer has appeared. The Universal Declaration of Human Rights, issued by the United Nations in 1968, states that men and women have a right to marry and found a family, but ignores the question of family size. The main omission, in my view, however, is the absence of any indication that rights imply obligations and that if one accepts the rights which follow from being a social animal, one must also accept the obligations. Among these, under present conditions, I would give a high place to an obligation not to produce children in demographically excessive numbers and not to take a substantial risk of begetting a defective child. But in the same way as rights imply obligations, obligations imply further rights. Thus, if couples have an obligation to restrict their fertility, they have a right to be fully informed of the methods available for doing so. Similarly, if they have an obligation to avoid producing defective children, they have a right to the best possible genetic counselling, to pre-natal diagnosis and, if necessary, to selective abortion.

The idea of quantity control for the family in the form of indiscriminate contraception, and in many countries indis-criminate abortion, is now widely accepted in principle, and, if I may extend the industrial simile, the idea of quality control should follow. In saying this, I am not thinking of genetic engineering in the sense of producing supermen or standard men by chromosome manipulation; I am thinking only of the elimination, so far as possible, of mental or physi-cal defectives. But even negative quality control of this kind is going to take a long time to become generally practicable and acceptable. In the meantime, we have the urgent need to extend quantity control from the family to the human race, from planned parenthood to planned population, and sooner or later some consideration for demographic problems will have to be injected into the traditional ideas of family planning. Nations will have to evolve population policies based on global rather than on national considerations. The mutual dependence on each other of the peoples of the world increases year by year, and no individual country should now stand aside in demographic isolation. We have only one world in which to live and in the long run we shall sink or swim together.

8

Sexuality and reproduction

Reproduction in man, though cumbersome, prolonged, and somewhat messy, is comparatively straightforward; certainly man has avoided the reproductive eccentricities found in many other species. In man, fortunately, we do not find delayed implantations, regular embryo-splitting, multi-ovulation, functional one-sidedness of the reproductive organs, or prolonged survival of spermatozoa in the female tract.

Nevertheless, man, in common with monkeys, has special problems deriving from the very limited reproductive capacity of the female, combined with the spread of her sexual receptivity throughout the menstrual cycle and the almost unlimited potential of the male. The result, even ignoring sexual outlets other than coitus, the only one that can lead to reproduction, and without the intervention of contraception, suggests a distinction between reproduction and sexuality. In some ways, however, there may be more connection than is generally realized. For instance, it has been suggested[21] that frequent coitus, by ensuring that a large number of sperm are usually present in the uterus, will facilitate the fertilization of both eggs in a twin ovulation, and therefore increase the likelihood of a twin birth. And there are other ways in which coital rates, a quantitative index of sexuality, could affect reproduction qualitatively. Take, for instance, our old friend the sex ratio.

The sex ratio

One of the solid facts of human reproduction is that, almost everywhere in the world, more males are born than females.

The excess is not great—of the order of 106 males to 100 females—and it varies slightly from time to time and place to place, but on the enormous numbers of births available for study, the excess is highly significant. Assuming that X and Y spermatozoa are produced in equal numbers, how does this disparity come about? Does pregnancy wastage, for instance, fall more heavily on the females? On the contrary, the evidence is that wastage during pregnancy, as later in life, falls more heavily on the males, so that the sex ratio must be higher at conception than at birth.

This conclusion is not easy to reconcile with the X and Y chromosome mechanism (see also p. 87). Do the Y spermatozoa survive better the process of maturation in the male tract, or have they some advantage of vigour or survival in the female tract? Such an idea may sound fanciful at first, but it becomes less so when we consider the hazards that spermatozoa meet in the course of traversing, apparently by random scatter, the cervical mucus, the uterine lumen, and the Fallopian tube. In these circumstances, some very small advantage might ensure the arrival at the egg of an excess of Y spermatozoa or facilitate their penetration of the egg.

Delayed fertilization

But there are difficulties in assuming that Y spermatozoa have an advantage in the female tract. It seems to be accepted that delayed fertilization, that is, penetration of the egg by a stale sperm, or vice versa, may result in an increase in foetal abnormality. Thus the well-known parental-age effect on the incidence of abnormality in the offspring may be due, not only to the increasing age of the oocyte population in the ovary, but also to decreased coital frequency in the older age groups. Certainly, this would account for the rather mysterious effect of paternal age on the abnormality rate, less than that of maternal age (see p. 9) but nevertheless distinct.

How, if at all, does this idea bear on the problems of the sex ratio? More than a century ago it was postulated, as a result of observations on farm animals, that the sex of the offspring depended on the state of the ovum at the time of fertilization, perfect ripe eggs producing males and imperfect overripe eggs producing females. The philosophy behind this idea was that only the best was good enough for males, anything would do for females. The comparatively modern phase of this saga began with a study of the leave-periods of German soldiers during the First World War in relation to their wives' last menstrual periods and the sex of the offspring. The author concluded that insemination early in the cycle resulted in an excess of boys and later in the cycle in an excess of girls. The credibility of this research was soon questioned, apparently even by the author himself, but the idea dies hard and it has recently been revived in the United Kingdom with unfortunate publicity of the 'more love, more boys' variety, a slogan which might bear hardly on a woman who has not produced an heir.

In the light of modern knowledge, what are we to make of this sort of stuff? It is credible that fertilization of an old egg on the verge of functional disintegration could lead to foetal abnormality, but it is difficult to see how this can be applied to the sex-ratio problem. Without being an exponent of the more extreme forms of Women's Lib, I am sufficient of a feminist to refuse to believe that a woman is a woman because she was derived from a stale egg. One obvious possibility is that the ratio of X to Y spermatozoa in the female tract, due to differential survival, changes with time after insemination.

From the sex ratio at birth and probably at conception, it would seem that the Y spermatozoa have on the average some initial advantage of numbers or vigour, but what if this advantage were abolished during ageing of the spermatozoa in the female tract, that is to say, if the Y spermatozoa had the speed and X spermatozoa the stamina? Then insemination shortly before the time of ovulation would result in an excess of boys, because fertilization would probably be by a younger spermatozoon, and earlier insemination in an excess of girls, because fertilization would probably be by an older spermatozoon. I have heard it said that this *must* be so because older and therefore wiser sperm would naturally produce females.

Frequency of intercourse

From this we can proceed to the self-evident proposition that the more frequent the insemination, that is, the greater the frequency of coitus, the more likely are the eggs to be fertilized by younger spermatozoa, and thence to the idea that increased coital frequency leads to a higher proportion of male offspring. This could explain the higher proportion of males born to young couples and early in marriage, when coital rates can be expected to be higher, and to the higher proportion of males said to be produced by artificial insemination, for which several inseminations are made around the expected time of ovulation. Could it also explain the slight seasonal variation in the sex ratio at birth and—presumably, though not necessarily—at conception? Is the clear seasonal variation in the birth-rate and sexual activity (see p. 54) related in any way to seasonal variations in the sex ratio? There is such variation in the sex ratio (Fig. 8.1) but in the United States at least it does not correspond with variations in the conception-rate or in the birth-rate.

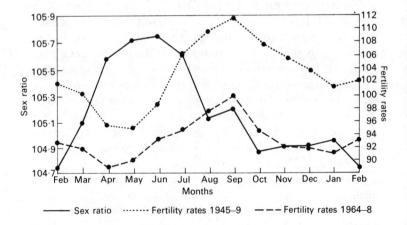

FIG. 8.1.   Fertility and the sex ratio in the United States.[27] Variation in the sex ratio does not correspond to variations in either the conception-rate or the birth-rate.

The significance of ideas regarding the age of spermatozoa clearly depends on the frequency of coitus in relation to the survival time of the ova and spermatozoa. For instance, if the functional survival of spermatozoa is limited to, say, 2 days, probably an overestimate, then intercourse every other day at least would be required to ensure that functional spermatozoa were always present at the site of fertilization. Frequency of intercourse varies with age and the duration of marriage, but according to a recent summary of a great deal of the available information, intercourse may average around 2·0–2·5 times per week, a rate marginal for maximum reproductive efficiency if equally distributed through the menstrual cycle.

Concentration of coitus around mid-cycle would, of course, reduce the total number of coitions per cycle required to effect fertilization and concentration at other times would reduce the chance of fertilization. Obviously, increase in the frequency of intercourse would sooner or later run into the law of diminishing returns, but taking 50–70 million sperm as an average daily production in young men, it would seem that intercourse twice a week or even every other day would be compatible with the maintenance of a sperm-count well above that normally regarded as marginal for fertility.

Distribution of coitus during the menstrual cycle

What, then do we know of the distribution of coitus within the menstrual cycle? This is a question of especial interest to the reproductive biologist, accustomed to the fact that, in mammals generally, coitus is normally restricted to the period of oestrus, and in many species is impossible at other times because of the development of a vaginal closure membrane. Coital activity throughout the cycle is restricted to primates, and even in monkeys tends to be greatest during and immediately before the expected time of ovulation (Fig. 8.2).

Detailed experiments[32] with monkeys show clearly that the increased attractiveness of the female at this time is olfactory in nature and is due to odorous sex attractants produced by the vaginal epithelium under the influence of oestrogen. These attractants can be extracted from vaginal washings and used to make attractive otherwise unattractive ovariectomized females, which could also be made attractive by the injection of oestrogen. Using oestrogenized ovariectomized females as a source of material, Michael and his colleagues showed that the operative substances were aliphatic acids, the mixture of which could be prepared artificially to give the same result. The effect was not species-specific, and extracts of human vaginal washings made ovariectomized monkeys attractive to males.

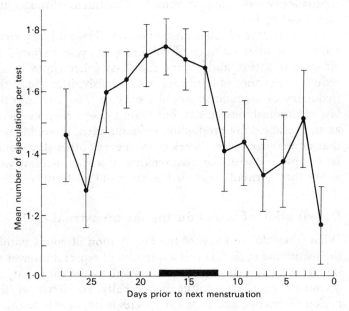

FIG. 8.2.   Copulatory activity of rhesus monkeys by reverse cycle days.[32] There is peak activity at mid-cycle, corresponding to the time of ovulation. Vertical bars indicate standard deviation.

What light does this work throw on human sexual behaviour? Even if intercourse occurs throughout the cycle, it might be expected, on biological grounds, to be more frequent at mid-cycle around the time of ovulation. Marie

Stopes, in her book *Married love*, published in 1918, gives diagrams indicating that 'natural desire' in a woman is greatest at mid-cycle, and pre-menstrually, but this was based on subjective impressions and in any case would not necessarily indicate the incidence of intercourse. Accurate information of this kind for man is naturally difficult to get, especially as recollections or general impressions are of little value and methods of analysis may be misleading. There is, in fact, no agreement; those who have investigated the problem variously claim that the greatest frequency occurs post-menstrually, at mid-cycle, or pre-menstrually. Both the available prospective studies[29, 42] based on contemporaneous daily records suggest a peak in the second week of the cycle (Fig. 8.3), a result entirely in accord with the work on monkeys. Whether this could be due in women to the pro-

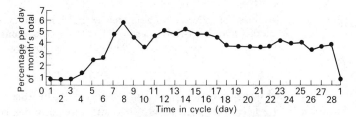

FIG. 8.3.   Distribution of coitus during the menstrual cycle.[29] As with the rhesus monkey there is peak activity about the time of ovulation. 1123 coitions recorded by 52 women.

duction of vaginal sex attractants, as it appears to be in monkeys, is perhaps open to question. It is not likely that olfactory pheromones of vaginal origin, assuming that they exist in man, would survive the present obsession with baths and deodorants. Moreover, psychological and social conditions might well obscure exocrinological effects. Nevertheless, the work on monkeys[32] is highly suggestive for man, and it is perhaps significant that the concentration of coitus shortly before the expected time of ovulation has been found to be restricted to what were described as 'lower-class' women, who might perhaps be more overtly pheromonal (Fig. 8.4).

Nor can the possible existence of sex attractants in the male be ignored, however slight the evidence.*

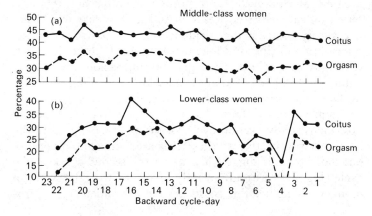

FIG. 8.4.   Distribution of coitus and orgasm during the menstrual cycle in middle- and lower-class women.[42] The lower-class women have a peak at mid-cycle; in middle-class women the peak is earlier.

Such considerations may account for some of the discrepancies in the literature. An attempt to summarize work on this subject revealed that of twenty authorities surveyed, nineteen recognized a post-menstrual peak of sexual desire, and about one-half claimed that there was a lesser or greater pre-menstrual peak. In all such discussion, however, it must be recognized that a post-menstrual peak might have no more physiological basis than a feeling of release on the part of the woman and of break-fast on the part of the man. Similarly, an increase in frequency towards the end of the cycle might be due to trying to relieve pre-menstrual tension, to reconciliation following an outburst of pre-menstrual

* The following story was recently culled by *The Biologist* (the journal of the Institute of Biology) from a national United Kingdom newspaper:
A Wigan teacher is reported to have sent a boy home because of his smell. The boy's mother sent her this note. 'Dear Miss, our Johnny smells the same as his dad and his dad smells lovely. I should know I've slept with him for 25 years. The trouble with you, Miss, is that you're an old maid and don't know what a proper man smells like.'

irritability, or even to pending menstruation inducing a feeling of 'do it now'. Until extensive and fully controlled experiments can be carried out, if ever, it is virtually impossible to separate the physiological, psychological, and social factors involved.

Can ovulation be accelerated by coitus?

There is a further complication in discussing the relation between sexual desire, incidence of coitus, and the time of ovulation—the possibility that ovulation may be accelerated by coitus. There is no doubt that ovulation occurs spontaneously in women, regardless of sexual activity, so that there is no valid analogy with animals such as the rabbit and the ferret in which ovulation is dependent on mating. Again, few would maintain that a second ovulation could be caused by coitus after the spontaneous ovulation had occurred, though this has been suggested. On the other hand, the idea that a follicle due to ovulate in a few days might do so more rapidly as the result of intense sexual excitement is not inherently improbable and cases of isolated coitus indicate that conceptions may occur well before the expected onset of the fertile period (see Fig. 2.10).

In a recent survey it was noted that of eleven girls requesting a legal abortion, four maintained that the pregnancy dated from a single or isolated coitus, which is well above probability if the relation between coitus and conception had been a causal one. In one of the four girls the pregnancy was said to result from rape, which is reminiscent of the belief that rape has a fertility rate so high as to imply accelerated or even induced ovulation. An extensive survey of cases of rape came up with the figures given in Table 8.1. Two conclusions stand out from this table:

(1) that the fertility of rape during days 7–9 of the cycle is higher than would be expected if ovulation had occurred at the usual time;
(2) that the fertility rate during days 10–18 is far higher than would be expected from a single normal coition during this period.

TABLE 8.1

Pregnancy following rape

Time in cycle (days)	Total number of cases	Number conceiving	Number not conceiving
1–6	56	12	44
7–9	15	11	4
10–18	54	53	1
19–28	45	3	42

From G. Linzenmeier. Zur Frage der Empfängniszeit der Frau: Hat Knaus oder Stieve recht? *Zentbl. Gynaek.* **69**, 1108 (1947).

These figures are certainly compatible with the idea that rape can accelerate ovulation, but there are several possible criticisms, such as the query whether cases of rape resulting in pregnancy are more likely to come to light than non-fertile ones, and the curious distribution of cases through the cycle. In these data the greatest frequency of rape per cycle day is on days 1–6, and it is odd, to say the least, that girls should have the greatest chance of being raped during menstruation or just afterwards. However, if the results can be substantiated they are of considerable interest in suggesting that the anger and fear evoked by rape may have the same effect as intense sexual excitement is said to have in accelerating ovulation.

Quality of intercourse

This brings me to my next point. If we assume that sexual excitement may accelerate ovulation then we come up against the problem of the quality of intercourse, and the current argument about female orgasm. Not long ago, a sociologist surveying cases of high fertility in an industrial town in the United Kingdom said to one woman: 'I see that you have ten children, Mrs. X, you must enjoy intercourse very much', to which the woman replied 'Oh, no, I just lie like a fish on a slab.'

With the present intense publicity about sex education, contraception, etc., women such as Mrs. X are probably becoming rare, but this story illustrates two well-known facts: (a) that coitus means little to some women except the hazard of becoming pregnant; and (b) that female orgasm, or even interest in coitus, is not necessary for fertility, and this is of course confirmed by the efficiency of artificial insemination in women. But fertility is not an all-or-nothing phenomenon. Fully functional people can usually be fertile with any partner, though there are rare instances of physiological incompatibility where a woman is infertile with a particular fertile male, though fertile with other males. Where fertility is biologically marginal, however, social factors such as the frequency and quality of intercourse may be of primary importance. Coitus twice a month is less likely to effect fertilization of a single short-lived egg than is coitus say, three times a week, and one woman may evoke intercourse more frequently than another, even with the same man, and vice versa. The question of the effect of quality of intercourse is more difficult, but two factors may be considered.

First, if intense sexual excitement may accelerate ovulation (see p. 119), the coitus generating such excitement is more likely to result in conception than is one which has only a casual time relation with ovulation. Second, the vaginal environment, odd as it may seem, is hostile to spermatozoa, and a quick entry into the cervical mucus is necessary if the spermatozoa are to ascend the female tract. And here, female orgasm promoted by a good social relationship and happy intimacy of the couple, might tip the balance in favour of conception in a case where the sperm count was marginal for fertility. Conversely, a poor social relationship, leading to perfunctory intercourse and lack of female orgasm might tip the balance towards prolonged infertility.

In this connection observations recorded by means of intra-uterine pressure-sensitive radio pills, are of importance.[16] An intra-uterine pressure build-up has been detected during female orgasm followed by a sudden relaxation, which would have the effect of drawing into the uterus spermatozoa already deposited in the vagina, and the authors suggest that

this might be an important factor where the sperm-count is
low (Fig. 8.5).

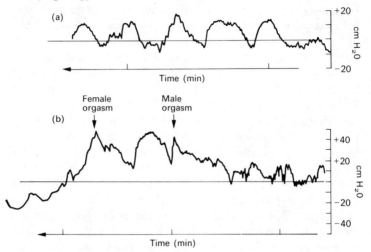

FIG. 8.5.   Intra-uterine pressure during coitus.[16] Pressure builds up and
relaxes suddenly after female orgasm.

Such a factor, however, could operate only after the deposi-
tion of semen in the vagina, so that multiple orgasm in the
female would not necessarily help. This point is well illustrated in the remarkable record buried in a book on the
pulse-rate and later resurrected (Fig. 8.6).

What can be said in conclusion? It is self-evident that,
within limits and other things being equal, frequency of
intercourse must affect the chances of conception occurring,
that is, it affects fertility. How far it can affect twinning,
the sex ratio, and similar aspects of human reproduction is
less clear. Obviously, experimental research by persuading
people to follow certain patterns of intercourse is probably
impossible on any adequate scale. Observational investigation, based on clear contemporary records of intercourse and
maternity outcomes should be possible, but the difficulties
are great.

Quality of intercourse obviously has little, if any, role in
effecting conception with fully fertile people. Its possible

role in cases of reduced fecundity, especially reduced sperm-count (see p. 121), offers a fascinating field of research, but the task of producing credible results might well daunt the most resolute investigator.

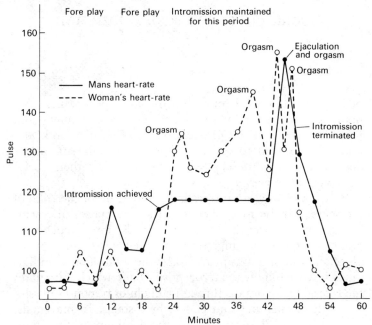

FIG. 8.6.   Effect of intercourse on the pulse-rate.[15]

9

Social effects on sexual function

Social contact between animals of the same species, including
man, is maintained by the chemical senses of smell, and
possibly taste, and the physical senses of sight, sound, and
touch. The extent to which individuals of different species
depend on the different senses for assessing their social
environment, however, varies enormously.

In a broad way, an interesting comparison of the roles of
the senses can be made between the two classes of warm-
blooded vertebrates, birds and mammals. Birds depend far
more on sight and sound than on smell, taste, and touch;
especially, the stimulus provided by the sight of other individuals is a well-known exteroceptive factor which stimulates
the reproductive organs through a chain of responses involving both the nervous and endocrine systems. In mammals,
by contrast, smell and touch are the dominant factors, both
being involved in the licking and nuzzling characteristic of
·courtship in many mammals; sight and sound are of
secondary importance in evoking sexual function except
where the individual has been conditioned to associating
particular visual or auditory stimuli with sexual activity.
Concordantly, there is enormous development of the olfactory tracts in many animals other than primates, in which
the convergence of the eyes to give stereoscopic vision has
restricted the space available for the nasal epithelium (Fig.
9.1). Nevertheless, man, even with his degenerate apparatus
can detect certain odorous substances in extraordinarily low
concentration (e.g. one-millionth of a milligram per cubic
metre of air).* The olfactory acuity of many lower mammals

* The minuteness of the quantities needed for olfaction is amazing. It has

can thus be more easily imagined than described, and rendering anosmic an animal which depends primarily on its sense of smell must be as disabling to it as is blinding to a human being.

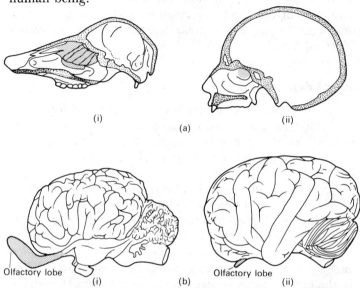

FIG. 9.1. (a) The area occupied by the olfactory epithelium of (i) the deer and (ii) man. (b) The olfactory lobes of the brain of the horse (i) and the gorilla (ii). In each case, size gives a good indication of the relative importance of the sense of smell in the two species.

Social contact, or lack of it, by whatever media, has three different types of effect on sexual function: imprinting of the young through neural channels; the rapid influencing of sexual behaviour in the adult; and the slow evocation of changes in the reproductive organs through neurohumoral channels. Of these, imprinting is usually permanent while the others are reversible.

Significant anomalies of the social environment include

been calculated that 1 g of muskone, the active principle of musk, in producing its strong smell would lose only 1 per cent of its weight in a million years.

deprivation of social stimulus (especially maternal deprivation), overcrowding, isolation (especially from the opposite
sex), and conditioning. Effects on sexual function operating
by these factors and through the channels mentioned are
many and diverse, and may result in non-appearance, cessation, retardation, acceleration, sensitization, or aberration
of sexual function.

Social deprivation

Some mammals show normal function and breed freely even
if isolated as a male and female pair from the onset of
puberty. The same must be supposed to apply among humans
in the unlikely event of a boy and girl being isolated together
for life.

Many other mammals, however, require the stimulus of a
social group for proper breeding performance and show
much impaired function if isolated as a pair. This has had an
interesting repercussion. In the past, many zoos have been
regarded as living museums which should contain one of
each sex of as many species as possible. Breeding performance
in these circumstances was often poor. Current thought, by
contrast, has veered towards the idea that zoo animals should
be kept in social groups with a structure characteristic of the
species in question, even if this means that fewer species can
be kept. Under such conditions, breeding performance has
much improved, although many difficulties arise from aberrations of behaviour resulting from the inevitably restricted
space. For instance, baby-stealing by the dominant female
has been observed in a group of breeding chimpanzees.

Many animals are brought into zoos or obtained as pets
when young, and may thus, especially in the case of rare
animals, be reared in isolation from other individuals of
their own species. Such animals may become 'humanized',
imprinted on their human attendant or on human beings in
general, and subsequently are indifferent, even though in
breeding condition, when introduced to other members of
their own species, including those of the opposite sex. Countless examples of this phenomenon could be cited, typified by

the motherless lamb, bottle-reared by the farmer's wife, which refuses later to join the flock.

A good example was seen in Chi-Chi, the giant panda brought to the London Zoo when very young and kept there, a thousand miles or so from the nearest other panda. When Chi-Chi became adult she showed intense and comparatively regular oestrous behaviour, which became directed towards her keeper or any other handy human. Sent on a visit to Moscow Zoo in the hope that she would mate with An-An, the male panda, she continued when in oestrus to make overtures to her human associates but actively resisted An-An's best efforts. A return match at the London Zoo had the same result.

As to the reverse attachment, it is unlikely that outside the realms of fiction and legend, any human being has been reared from a very early age entirely by individuals of a different species in such a way as to effect imprinting, so that the effects can only be conjectured. It would be an interesting experiment. In the meantime, every possible grade of relationship of humans to animals is to be seen, even to attempted sexual intercourse, which if Kinsey is right may be more common than is generally thought.

Fixation of an individual on one of another species does not, of course, always involve man; it may sometimes occur between individuals of different animal species. Thus Andean farmers in Peru are said to use vicuña males imprinted on alpacas to facilitate the otherwise difficult cross-breeding.

A rather different kind of deprivation, for instance, maternal deprivation, in certain primates where the mother–baby relationship is very strong, also has serious effects. Thus, if Rhesus monkeys are removed from the mother at birth, reared on cloth imitation mothers, and bottle-fed, many of them later show sexual aberrations, including disinterest in the opposite sex. Analogous effects of maternal deprivation are not uncommonly seen in man.

The acute effects of deprivation, in the sense of isolation, on the sexual function of baboons have been described. Removing a female from a group and placing her in isolation has the immediate effect of shortening the menstrual cycle by curtailment of the quiescent phase following menstrual bleeding. The process is reversed on returning the female to the group. The lengthening of the quiescent phase in grouped females can be regarded as being the result of social trauma, and it may be analogous with the disruption of the cycle seen in female mice kept in large groups without males (p. 134).

Social dominance and sexual preference

Any group of animals whether of hens or schoolboys soon sorts itself into a 'peck order' or social hierarchy and in a heterosexual group the dominant males monopolize the females. Conversely, the lower order males are denied access to females. It is now known that for the male this is an auto-catalytic process; dominance and copulation increase the flow of testosterone, which in turn increases dominance and sex-drive. By the reverse sequence of events, inferior males become more inferior because of the depression of testo-sterone levels. Associated with such social dominance is a comparative lack of sexual preference.

As a general rule, any male will serve any available female of the same species and any receptive female will take any male. There are, however, well-known exceptions, especially where one sex or the other is offered a choice. With the male, familiarity may not breed contempt, but it can certainly breed diminution of sex drive, and instances are well known where a male has refused to copulate with long-familiar females, or has done so only half-heartedly, but has mated enthusiastically with unfamiliar ones. Much of this information comes from individual case histories but the familiarity effect has recently been demonstrated very clearly in experiments with male ferrets, which are not usually thought of as lacking in sex drive, but which in the experiments cited copulated more vigorously and for longer periods with unfamiliar than with familiar females.*

* Environment may also play a part. A pair of green sea turtles showed no sexual activity while isolated for some months in a large concrete tank but immediately began to copulate when returned to the breeding pool and sub-

With the female, some extraordinary instances of sexual preferences have been demonstrated in mice. Given the choice, adult female mice prefer males of a strain other than their own. Even more fanciful, one author, Mainardi, working in Parma, showed that females reared as nestlings in the presence of the father preferred males of a strain different from that of the father when they become adult. Reared without the father, their taste in males was more catholic. This was obviously a case of olfactory imprinting, because the reaction was not seen when the smell of the father was blocked by a non-specific odour, appropriately, in these particular experiments, Parma Violet.

The possible significance for man of such instances of sexual preference offers a fascinating field of speculation. There can be little doubt of the existence of a familiarity effect in man. The decrease in the coital rate as a marriage proceeds is not entirely a matter of ageing—a change of partner is credibly said to result in a return to greater coital frequency.

Can the Mainardi effect explain why so many daughters marry men so very different from their fathers? How in fact do human beings choose their mates? The question, perhaps, is not as difficult as might be supposed. In some cultures, the answer is simply that their parents do it for them, and that prospective husband and wife do not meet until the wedding day, or not even then where marriage by proxy is permitted. At first sight this may seem to be a deplorable system, but it can be argued that parents, with their greater knowledge of the world and of people, are likely to make a better choice than their relatively inexperienced offspring, and this is especially true where early marriage is customary.

In other cultures, those concerned again have little or no say in the matter, their choice being determined by tribal

---

jected to a different environment and the competition of other turtles. Not very dissimilar is the advice given by Masters and Johnson that busy but sexually inadequate couples should take a holiday in a luxury hotel and follow a course of treatment. A review of the book in the *Journal of Biosocial Science* pointed out that living in a luxury hotel with nothing to do except make love might well render other treatment superfluous.

custom, of which they are merely pawns. Alternatively, it may be only the females who have little choice, the most attractive females being annexed without any option by the dominant males, leaving the residue to the subordinate males.

It is, however, in those parts of the Western world where both sexes have theoretically a free choice that the most curious manifestations of dominance and selection are to be seen. One may indeed wonder how any combination of the two, plus an admixture of that intangible quality known as sex appeal, resulted in the strangely assorted couples to be seen around. In reality, the choice is of course limited by the number of social contacts, long-term or casual, made by any particular individual before marriage. In many cases the number of contacts may be relatively small and therefore unlikely, allowing for all the variables involved, to result in the meeting of a man and woman ideally suited to each other. Few people would rely on haphazard social contacts for filling any other sort of vacancy; the field would be surveyed by advertising or by consultation with an employment agency. This is the justification for the existence of marriage bureaux, and it is to be hoped that greater use will be made of these increasingly reputable organizations.

And there are other methods of surveying the market. Indian newspapers, even today, carry columns of advertisements for spouses, often with detailed specifications. No doubt some of these are inserted by parents, but many appear to come from the individuals themselves, among which Ph.D.s approaching the age of 30 years are prominent. Here are two specimens from *The Times of India*:

H——— K——— Pandit University Lecturer, 29 belonging to tradi-tionally respectable family, earning Rs. 800, needs a very beautiful intelligent highly educated Kashmiri Pandit bride with good voice and of noble lineage.

Wanted Suitable Match For Agarwal girl. The girl is 21 years of age, 152 cm. tall, slim, fair & beautiful. She is a qualified beau-tician and lives with her parents in England where they have a well established business. The boy should be professionally qualified and be prepared to marry in England. In this connection, permission to immigrate can be obtained.

Social factors in sexual behaviour

Many features of sexual behaviour and response appear to be independent of social stimuli or even previous experience; these are irrelevant to the present theme. Much of sexual behaviour, however, is conditioned by the presence or near presence of the opposite sex. In mammals, there is nothing as dramatic as the olfactory sex attractants of many insects which enable a male to track down a female from a distance, often a great distance, but smell nevertheless plays a large part in mammals in the recognition of sex and sexual condition and therefore in sexual behaviour.* Sight and sound play minor parts. Such effects on sexual response are mediated directly through the central nervous system and appear rapidly. The previous experience of the individual may also be important. Three examples may be given.

The first relates to the immobilization reflex which appears in the oestrous sow and immobilizes the animal during coitus, which is very prolonged in this species. This reflex is elicited in oestrous sows by the mounting of the boar. In the absence of the boar, pressure on the back similar to the mounting pressure evokes the response in only about 50 per cent of oestrous sows. Analysis of this influence of the boar was carried out by (a) making a tape recording of his grunts and playing it back to the sow and (b) putting the sow into a pen recently vacated by a boar. In either case 70–80 per cent of oestrous sows responded to pressure on the back—the two stimuli together produced a response in 90 per cent.

The immobilization reflex depends, therefore, on both the sound and smell of the boar. In the absence of the boar, exposure to the smell of his semen, urine, or saliva elicits the

* On a visit to a well-known perfumery near the London docks, I was told that a group of girls engaged in the preparation of a particularly alluring brew had recently protested because, they said, every night when they left the factory they were followed by sailors.

reflex in many oestrous sows. Conversely, castration, which abolishes the boar's smell, abolishes his ability to evoke the reflex. It would seem, therefore, that some odorous substances produced by the testis, or some metabolite of a testicular substance, is responsible for this remarkable effect of the boar in completing the oestrous syndrome of the sow. This substance is now known to be androstenone, an odorous steroid related to testosterone. It also appears that, being present in the circulating blood, this substance escapes at any point of external secretion through the kidneys, seminiferous tubules, accessory sex glands, or salivary glands. The immobilization reflex in the sow may be likened to the exaggerated curvature of the spine (lordosis) commonly displayed by oestrous females in the presence of a male, as in the domestic cat.

The second example relates to the evocation of sexual behaviour in stallions on approaching the oestrous mare or a dummy. From Fig. 9.2 it will be seen that the experienced adult stallion is somewhat more responsive to the mare than is the young one, and that neither's reactions are affected by blindfolding, which leaves smell and sound as the operative stimuli. Smell is obviously the dominant factor, but sight also plays a part with the experienced stallion, which responds fairly well to the dummy, as opposed to the young stallion which does not respond to it. The mature stallion's behaviour suggests that we have here a case of a neural sensory response conditioned by experience.

In men, the relative contributions of innate reactions, the presence of the opposite sex, and the extent of pre-conditioning vary enormously from individual to individual. Penile erection, for instance, is certainly not primarily dependent on social stimulus or conditioning, but it can be intensified by both, as shown by the work of Pavlov on eliciting spontaneous erection in dogs as a conditioned reflex evoked by odour.

In the human female the innate component of sexual behaviour is less strong, and sexual excitement with flow of secretion from the vaginal epithelium, spontaneous copulatory movements, and orgasm may almost be called a learned response. According to Kinsey many women attain orgasm only after repeated intercourse over a period of time—and in 10 per cent not at all—and this is greatly influenced by a social factor: the compatibility, consideration, and understanding of her partner. This reaction of the female may be important where fertility is marginal (see p. 121).

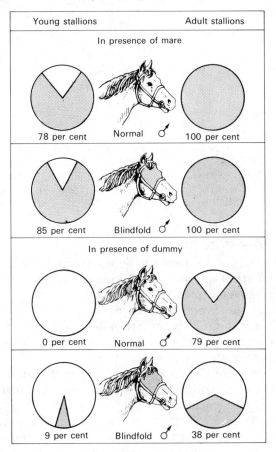

FIG. 9.2.   Evocation of sexual behaviour in the stallion.[17] The experienced adult stallion responds more readily to the presence of the mare than does the young one, but even blindfolded a high percentage of both respond. By contrast, only the adult stallion responds to the presence of a dummy and this response is largely eliminated by blindfolding.

Fig. 9.3 shows the uterine contractions of a cow in response to sexual stimulation. The height of a peak on the curve indicates the degree of contraction. Similar contractions of the uterus have been recorded in women during coitus and particularly at orgasm (see p. 122).

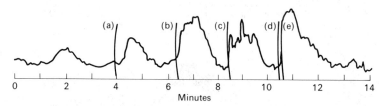

FIG. 9.3.   Response of the uterus of the cow to sexual stimulation.[43] Contractions of the uterus as measured by an intra-uterine balloon. (a) Bull brought into sight of cow. (b) Bull allowed to nuzzle vulva and hindquarters of cow. (c) Bull allowed to mount but not copulate. (d) Bull allowed to mount again. (e) Bull copulated.

Social effects on the sexual cycle

Work over the last decade has uncovered some very remarkable effects of the social environment on the oestrous cycle and pregnancy, mediated apparently by olfactory stimuli affecting the hypothalamus, the anterior pituitary gland, and thence the gonads. Such stimuli act slowly. For instance, it has been reported that the introduction of males into a group of female goats or sheep results in peak mating some days later, and this appears to be a case of partial synchronization of oestrus following the stimulus provided by the presence of the males.

However, the most detailed information, and the only conclusive evidence that the smell of the male is the operative factor, has been derived from work on laboratory mice. Female mice, isolated in small numbers away from males, have oestrous cycles longer and less regular than those of females kept normally. When large numbers of females are kept together the cycles become very irregular or disappear (the Lee–Boot effect), but the introduction of males restores the cycles and results in a peak of oestrus 3 days later (the Whitten effect). An even more remarkable effect in mice is that pregnancy is blocked in a newly mated female by exposure to a male of a strain different from that of the stud male (the Bruce effect). Smell is known to be responsible for all these reactions in mice.

Does this work on social effects on the sexual cycle have any significance for man? The evidence so far is slight and mainly negative. It has been recorded that the age of menarche in girls is not influenced by whether they attend co-educational or one-sex schools, nor is there evidence that the regularity of the menstrual cycle is influenced by the presence or absence of males. As to the Bruce effect in man, the only indications known to me come from what may euphemistically be called folklore. Two instances in man may however be cited, one of which is certainly a social effect and the other possibly so. It has been found that in two-child families the second child menstruates earlier the longer the interval between births (Fig. 9.4). This could be because the second child was better cared for when the inter-birth interval was longer—an

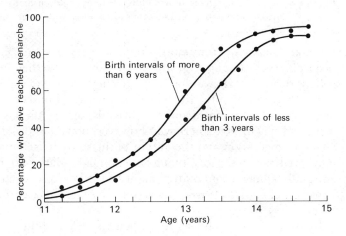

FIG. 9.4. The inter-birth interval and the age of menarche of the second child in a two-child family.[14] Menarche is about 4 months earlier when the birth interval is increased from less than 3 years to more than 6 years.

explanation which accords with the fact that it was also found that an only child experienced menarche earlier by 6 months or so than one of a large family (Fig. 9.5). It is possible but not easy to think of olfactory effects in this connection. An effect in man more likely to be olfactory in nature is

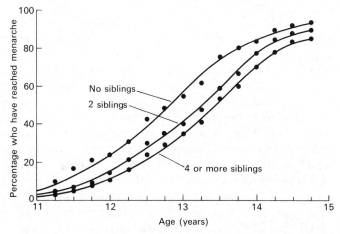

FIG. 9.5. Age of menarche of only children.[14] Only children first menstruate about 6 months earlier than girls with 4 or more siblings.

the menstrual synchrony and suppression reported to have been observed among a group of girls living together in a college dormitory. It was found, with a high degree of significance, that in 6 months the scatter in menstrual onset times decreased in room mates and close friends, that is, there was increased synchrony. There was, however, no actual proof of olfactory mediation and the author of the study herself gives some credence to the idea that a common light–dark pattern, especially among room mates, might be the operative factor; however, close friends were more nearly synchronized than room mates.

In spite of the doubt as to its general applicability, the work on the presence of other individuals as an external factor causing profound effects through the slow-acting brain–gonadal axis has great biological interest. The externally-secreted odorous substances concerned serve to integrate a population of individuals in the same way as internal secretions serve to integrate an individual, and they come under the comprehensive term 'chemical messengers', applied to hormones. The word 'pheromones', literally 'carriers of excitation' was coined to describe these external secretions acting on other individuals, as opposed to hormones which act within the individual.

The study of pheromones of the slow-acting type is comparatively new, and further exciting developments may be expected. In the meantime, it may be noted that androstenol, a compound closely related to androstenone, the sow-immobilizing pheromone of the boar, and also having a pleasant musk-like smell, is found in human urine.

Phantom pregnancy

A remarkable example of the effect of social pressures combined with maternal urges is to be seen in pseudocyesis, phantom pregnancy in women. This condition is characterized by cessation of menstruation, distention of the abdomen, enlargement of the breasts, and other stigmata of pregnancy which appear as a result of psychosomatic effects induced by an intense desire to become pregnant.

Phantom pregnancy has played a part in history, as witness the cases of Joanna Southcott, an English religious fanatic of the nineteenth century with 100 000 followers, who when more than 60 years of age claimed to be about to produce a new Messiah and was declared pregnant by doctors. More important was the case of Queen Mary (Bloody Queen Mary), daughter of Henry VIII and Catherine of Aragon who, married to Phillip II of Spain, had two episodes of pseudocyesis, during the first of which she notified the Doge of Venice of the birth of a son, but no pregnancy. This frustration to the idea of uniting the English and Spanish thrones by marriage could be said to have resulted in the Spanish Armada.

Pseudocyesis must be clearly distinguished from the condition of pseudopregnancy, a purely physiological condition seen in certain animals in which the corpus luteum develops after ovulation, even in the absence of fertilized eggs, to an extent causing development of the uterus and mammary glands reminiscent of that seen in early pregnancy. Such a condition of pseudopregnancy is seen in women in a rudimentary form in the luteal phase.

Overcrowding and sexual function

It is well known that with many animals an upsurge of population is followed by a catastrophic decrease which is not obviously attributable to the classical limiting factors of disease, predators, and food shortage. The significance of such occurrences for the human race is painfully obvious at the present time and justifies a close analysis of the phenomenon under both natural and experimental conditions. Several most illuminating laboratory experiments have been carried out. In one, rats were maintained at a density double that compatible with normal breeding, and extraordinary deviations of maternal and sexual behaviour were observed, as shown by the following quotations from a description of this work:

The females that lived in the densely populated middle pens became progressively less adapted to building adequate nests, and eventually stopped building them at all . . .
The subordinate males in all pens adopted the habit of rising early. This enabled them to eat and drink in peace . . .
Among the subdominant males there was much abnormal behavior. For instance, there was a group of homosexuals. They were really pan-sexual animals, and apparently could not discriminate between sex partners . . .
Another type of male merged in the crowded pens. This was essentially a very passive type that moved a good deal like a somnambulist . . .

The strangest of all of the abnormal male types described by Calhoun were what he called the 'probers'. These animals took no part at all in the status struggle. Nevertheless, they were most active of all the males in the experimental population, and they persisted in their activities in spite of attacks by the dominant animals. In addition to being hyperactive, the probers were hypersexual, and in time many of them became cannibalistic. They were always chasing females about, entering their nests, having intercourse with them in the nest—a thing that the normal rats would never do. These probers conducted their pursuits of oestrus females in a very abnormal manner, abandoning all courtship ritual.

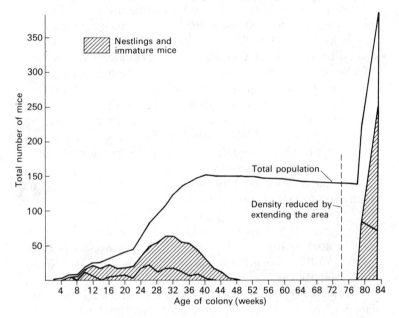

FIG. 9.6. Overcrowding in a mouse colony.[9] Breeding stopped when the population had increased sufficiently to cause overcrowding and started again when the density was reduced by increasing the space.

These results of overcrowding, suggesting a not very attractive prospect for man, appeared to be mediated by, or were at least associated with, an excessive number of contacts. Under these conditions breeding would be greatly reduced; we have other evidence of this effect. In a mouse colony kept in a restricted area numbers increased to a plateau and then stabilized because of the great reduction in the breeding performance of the females, few of which became pregnant, apparently because of the suppression of the oestrous cycle.

This may be thought of as a super-Lee–Boot effect (see p. 134), the dense overcrowding preventing the males from exerting their usual regulatory role. Sudden increase of the space available caused a sudden resumption of breeding.

A member of the Human Reproduction Unit at WHO carried out somewhat similar experiments in two enclosures of 13·3 square feet and noted that had the mice continued to breed as they did at the beginning of the experiment there would have been 100 000 mice in each enclosure by the end of a year. As it was the populations were held in check by a combination of decreased fertility and neonatal mortality. Maximum increase occurred during the third month, but there was already decreased productivity per adult female, with pregnancy rates falling steadily. Testes and spermatozoa of the males were normal, but a large proportion of males showed little sexual activity.

It is natural to speak of auto-regulatory mechanisms in such circumstances, but of what do these mechanisms consist? Here we have some guidance from other experiments. Continuous disturbance, especially of nursing mothers and their young, will naturally lead to excessive neo-natal mortality, but there are other factors, such as the olfactory effects described above.

These observations may have some relevance to man, faced with the current upsurge of population and the prospect of severe overcrowding in many parts of the world? Are self-regulatory mechanisms such as those we have noted likely to become operative, and if so at what point, and what will be the physiological pathways? Will the incessant ear-splitting noise which now seems inseparable from modern civilization decrease fertility in man as did auditory insults in one researcher's rats? And what other inhibitory factors might supervene?

The portents for the occurrence of these fertility-limiting effects are not hopeful. Women can and do become pregnant under conditions of gross overcrowding and savage stress, as witness the records of the Nazi concentration camps, and the reports of countless social workers. Here is one such report, made not many years ago by a domiciliary service worker operating from the Marie Stopes Memorial Foundation in London.

People still live in the most appalling conditions. The first family I ever visited lived in a tiny attic room. The husband was on night work and was in bed; three little boys romped around him as he slept; the mother sat on the bed feeding the baby where I joined her—there was nowhere else to sit. In that small room with the gas stove and all the impedimenta of daily living, the only play space for those children was the bed. In one house seven rooms are let to separate families. There are 33 children and 5 pregnant mothers . . . using one lavatory and one gas stove. One mother aged 34 neglects and badly treats her 14 children and is expecting a 15th child.

We may recall also the population explosion of the mid-nineteenth century in England which was associated with the crowding together of people in the slums of the industrial revolution, and the high birth-rate in the crowded tenements of the East and the shanty towns of South America.

So far as can be foretold, therefore, only the ultimate extremes of stress and overcrowding are likely to bring a self-regulating mechanism to human reproduction. Long before this stage is reached the world will be in even greater chaos than it is at present.

References

1. H. ALTNER. Location by the nose. In *Signals in the animal world*. George Allen & Unwin, London (1967).
2. K. BOJLEN and M. W. BENTZON. Eskimopigers Menarchealder. *Bibl. Laeg.* **156**, 109.
3. M. BONTE and J. VAN BALEN. Prolonged lactation and family spacing in Rwanda. *J. biosoc. Sci.* **1**, 97 (1969).
4. M. G. BULMER. The twinning rate in Europe and Africa. *Ann. hum. Genet.* **24**, 121; The number of human multiple births. *Ann. hum. Genet.* **22**, 158 (1960).
5. P. E. BROWN. Age at menarche. *Br. J. prev. soc. Med.* **20**, 9 (1966).
6. K. S. F. CHANG, S. T. CHAN, W. D. LOW, and C. K. NG. Climate and conception rates in Hong Kong. *Hum. Biol.* **35**, 366 (1963).
7. U. M. COWGILL. Recent variations in the season of birth in Puerto Rico. *Proc. natn. Acad. Sci.* **52**, 1149 (1964).
8. U. M. COWGILL. Season of birth in man. *Ecology* **47**, 614 (1966).
9. P. CROWCROFT. *Mice all over*. Foulis, London (1966).
10. K. DALTON. *The premenstrual syndrome*. Heinemann, London (1964).
11. Data of Registrar General's *Statistical Review of England and Wales*, Part III, Commentary (1959).
12. Data of Registrar General's *Statistical Review of England and Wales*, Part III, Commentary (1963).
13. G. K. DORING. Über die relative Haufigkeit des anovulatorischen Cyclus im Leben der Frau. *Arch. Gynaek.* **199**, 115 (1963).
14. J. W. B. DOUGLAS. The age of reaching puberty: some associated factors and some education implications. *Sci. Basis Med.* **6**, 91 (1966).
15. C. S. FORD and F. A. BEACH. *Patterns of sexual behaviour*. Eyre & Spottiswoode, London (1952).
16. C. A. FOX, H. S. WOLFF, and J. A. BAKER. Intra-uterine pressure during coitus. *J. Reprod. Fert.* **22**, 243 (1970).
17. E. S. E. HAFEZ, M. WILLIAMS, and S. WIERZBOWSKI. Extract from 'The behavior of horses'. In *The behaviour of domestic animals*, Chap. 13. Bailliere, Tindall & Cox, London (1962).
18. C. G. HARTMAN. *Science and the safe period*. Bailliere, Tindall & Cox, London (1962).

19. F. HEFNAWI, M. H. H. BADRAOUI, N. YOUNIS, and F. HASSIB. Lactation patterns in Egyptian women. I. Milk yield during the first year of lactation. *J. biosoc. Sci.* **4**, 397 (1972).

20. M. HENDERSON and J. KAY. Differences in duration of pregnancy. *Archs. envir. Hlth.* **14**, 904 (1967).

21. W. H. JAMES. Coital rates and dizygotic twinning. Secular changes in dizygotic twinning rates. *J. biosoc. Sci.* **4**, 101, 427 (1972).

22. O. JEANNERET and B. MACMAHON. Secular changes in rates of multiple births in the U.S. *Am. J. hum. Genet.* **14**, 410 (1962).

23. J. L. KING. Menstrual records and vaginal smears in a selected group of normal women. *Contr. Embryol.* **18**, 81 (1926).

24. A. C. KINSEY, W. B. POMEROY, and C. E. MARTIN. *Sexual behaviour in the human male.* Saunders, Philadelphia (1948).

25. A. LEFFINGWELL. *Illegitimacy and the influence of seasons upon conduct.* Swan Sonnenschein & Co., London (1892).

26. W. LEIDL. *Climate and sexual function in male domestic animals.* Hanover Schaper (1958).

27. W. T. LYSTER. Fertility and the sex ratio in the United States. *Am. J. Obstet. Gynec.* **110**, 1025 (1971).

28. N. MCARTHUR. Statistics of twin births in Italy. *Ann. Eugen.* **17**, 249 (1954).

29. A. A. MCCANCE, M. C. LUFF, and E. C. WIDDOWSON. Distribution of coitus during the menstrual cycle. *J. Hyg., Camb.* **37**, 571 (1952).

30. W. V. MACFARLANE and D. SPALDING. Seasonal conception rates in Australia. *Med. J. Aust.* **1**, 121 (1960).

31. S. MCKINLAY, M. JEFFREYS, and B. THOMPSON. An investigation of the age at menopause. *J. biosoc. Sci.* **4**, 161 (1972).

32. R. P. MICHAEL. Copulatory activity of rhesus monkeys by reverse cycle days. *Acta endocr.*, Suppl. 166 (1972).

33. C. M. MONGE. Demografía y altitud en el Peru, p. 32. In *Población y altitud* (ed. L. Sobrevilla *et al.*). Instituto de Investigaciones de la Altura, Universidad Peruana Cayetano Heredia (1965).

34. S. D. MORRISON. *Human milk.* Commonwealth Agricultural Bureaux, Farnham Royal, Bucks (1952).

35. H. B. NEWCOMBE and O. G. TAVENDALE. Maternal age and birth order correlations. *Mutat. Res.* **1**, 446 (1964).

36. A. S. PARKES. Seasonal variation in human sexual activity. In *Genetic and environmental influences on behavior* (ed. J. M. Thoday and A. S. Parkes). Oliver & Boyd, Edinburgh (1968).

37. A. S. PARKES. Environmental influences on human fertility. *J. biosoc. Sci.*, Suppl. 3, 13 (1971).

38. K. PEDERSEN-BJERGAARD and M. TØNNESEN. Sex hormone analyses. II. The excretion of sexual hormones by normal males, impotent males, polyarthritics and prostatics. *Acta med. scand.* **131**, Suppl. 213, 284 (1948).

39. Population Reference Bureau. Do Catholic countries have the highest birthrates? *Population profile, July* (1968).
40. Population Reference Bureau. Pockets of high fertility in the United States. *Popul. Bull., Wash.* **24**, 25 (1968).
41. J. THOMPSON. The growth phenomenon. In *The optimum population for Britain* (ed. L. R. Taylor). Academic Press, London (1970).
42. J. R. UDRY and N. M. MORRIS. Distribution of coitus and orgasm during the menstrual cycle in middle- and lower-class women. *Nature, Lond.* **220**, 593 (1968).
43. N. L. VANDEMARK and R. L. HAYS. Response of the uterus of the cow to sexual stimulation. *Iowa St. Coll. J. Sci.* **28**, 107 (1953).
44. WHO Technical Report No. 360. *Biology of fertility control by periodic abstinence* (1967).

Suggestions for further reading

F. A. BEACH. *Sex and behavior*. Wiley, New York.

R. B. GREENBLATT (ed.). *Ovulation*. Lippincott, Philadelphia (1966).

F. E. HYTTEN and I. LEITCH. *The physiology of human pregnancy*. Blackwell Scientific Publications, Oxford (1964).

A. C. KINSEY, W. B. POMEROY, and C. E. MARTIN. *Sexual behavior in the human male*. Saunders, Philadelphia (1953).

A. C. KINSEY, W. B. POMEROY, and C. E. MARTIN. *Sexual behavior in the human female*. Saunders, Philadelphia (1958).

W. H. MASTERS and V. E. JOHNSON. *Human sexual response*. J. & A. Churchill, London (1966).

A. S. PARKES. *Sex, science and society*. Oriel Press, Newcastle (1966).

A. S. PARKES (ed.). Biosocial aspects of human fertility. *J. biosoc. Sci.*, Suppl. 3 (1971).

J. M. TANNER. *Growth at adolescence*. Blackwell Scientific Publications, Oxford (1962).

For further information on the topics discussed in this book see the following articles written by the author.

Chapter 1. The reproductive life cycle. *Sci. J.* **6**, 26 (1970).

Chapter 3. Seasonal variation in human sexual activity. In *Genetic and environmental influences on behaviour*. Oliver & Boyd, Edinburgh (1968).

Chapter 4. Environmental influences on human fertility. In Biosocial aspects of human fertility. *J. biosoc. Sci.* **13**, Suppl. 3 (1970).

Chapter 5. Multiple births in man. *J. Reprod. Fert.* **105**, Suppl. 6 (1969).

Chapter 6. The sex-ratio in man. The biology of sex, *Penguin science survey, 1967* (ed. A. Allison). Penguin Books, Harmondsworth.

Chapter 7. The social impact of human reproduction. *J. biosoc. Sci.* **5**, 195 (1973).

Chapter 8. Sexuality and reproduction. *Perspect. Biol. Med.* **17**, 399 (1974).

Chapter 9. Social effects on sexual function. *Impact Sci. Soc.* **18**, 273 (1968).

Index

Fallopian tube, 2, 112
fertility,
  onset at puberty, 18
  decline at menopause, 20
  during lactation, 41
  of old men, 16
  demographic, 66
  examples of extreme, 101–3
  age-specific rates, 102
fertilization,
  determination of sex at, 87
  late, 112, 113
foetus, 9
follicles, 8
  see also ovary
gestation; see pregnancy
gonad; see testis, ovary
gonadotrophin; see pituitary, 10
Hutterites, high fertility of, 102
illegitimacy, 53, 57, 58
insemination,
  artificial, 121
  time of during cycle, 113
intercourse; see coitus
labia, 2, 6
lactation,
  attempts to stimulate, 37
  duration of, 37
  milk yield, 37, 38
  lactation curve, 38
  composition of human milk, 39
  fertility during, 41, 70
life cycle, 1
  reproductive phase of, 1
light, 62, 63
mammary glands, 2
  development of, 6
man;
  gregarious animal, 100
  reproductive capacity, 101
marriage patterns, 68
maturation;
  male, erection, 5
    orgasm, 5
    ejaculation, 5
  female, menstruation, 6, 7
    ovulation, 7
    see also menarche
menarche, 6, 7
  and ovulation, 7, 18
menopause,
  and fertility, 8, 9

characteristics of, 10, 11
age at, 17–20
  and birth interval, 136
  and number of sibs, 137
menstrual cycle,
  length of, 22
  variation in individuals, 22
  variation between individuals, 23
  anovular cycles, 8, 20, 29
  length of phases, 24
  side-effects during, 30
menstruation, 22
multiple births,
  see also twins, triplets, and quadruplets, 74
  sex combinations in, 75, 76
  ethnic variation in frequency and composition, 77–9
  relation with maternal age, 80
  relation with birth order, 80, 81
nutrition, 64
oestrogen, 4
olfactory effects,
  stimulus to coitus, 115, 116
  sex attraction, 131
  on sexual cycles, 135, 136
ovary, 7
  hormone production, 2, 7
  dual function of, 8
  follicles, 8
  oocyte population, 8
  ovulation, 9
ova,
  numbers of, 9
  age of, 9
  condition of in female tract, 9
overcrowding,
  effects on sexual cycle, 134
  and sexual function, 138, 139
  and population growth, 140, 141
ovulation,
  and menarche, 6, 18
  and menopause, 20
  during lactation, 41
  multiple, 74
  detection of, 25, 26
  acceleration of by coitus, 119
peck order;
  see social dominance, 128
penis, 2, 4
pheromones, 137

pituitary,
  gland, 5
  gonadotrophin, 10
population,
  growth of human, 101, 102
  growth of animal, 138, 139
pregnancy, 31
  length of, 31
  differences in length of, 32, 34
  length of with twins, 35
  phantom, 137
  pseudo-, 137
  block to, 135
pregnancy wastage,
  amount of, 36
  effect on sex ratio, 90, 91
prostate gland, 2
puberty,
  in boys, 5, 16
  in girls, 7, 17
quadruplets,
  frequency of, 75
  sex combinations in, 75
rape,
  pregnancy following, 120
  incidence during menstrual
    cycle, 120
reproductive function, variation in,
  individuals, 14
  ethnic groups, 15
  different climates, 14
reproductive organs, 1, 2
scrotum, 2, 59
seasonal variation,
  in birth rate, 44, 49
  in conception rate, 46
  in sexual activity, 54
  effect of length of daylight, 44
  effect of temperature and humi-
    dity, 50, 51
  the comfort factor, 47
  the festive seasons, 52
  in illegitimacy rate, 53, 57, 58
secondary sexual characters, 3, 4
seminal vesicles, 2
sex ratio, 112
  chromosome mechanism, 87
  affected by mortality during
    gestation, 88
  at conception, 89
  in old age, 89
  effect of postnatal mortality, 89

during pre-natal life, 90
  at birth, 93
  ethnic variation, 94
  secular variation, 95
  effect of parental age, 96
  effect of birth order, 96, 97
  illegitimacy, 98
  during the life cycle, 98, 99
  time of insemination in cycle,
    115
  and frequency of coitus, 112
sexual behaviour, 131
sexual desire, in women, 117
sexual preferences, 127, 128, 130
sexuality, 11, 12, 111
Siamese twins, 74
social deprivation, 126
social dominance, 128
social effects on sexual function,
  124
social factors, 66–9
social hierarchy;
  see social dominance, 128
spermatogenesis, 5
spermatozoa, 2
  numbers of, 115
  survival of in female tract, 113,
    115
temperature, 59
testis, 4
  dual function of, 4
  spermatogenesis, 5
  hormone production, 2, 4, 5
  size changes, 4
  descent of, 4
testosterone, 2
triplets,
  frequency of and sex combina-
    tions in, 75
twins,
  and frequency of intercourse, 9
  effect on family size and birth
    interval, 71, 72
  frequency of, 74
  dizygotic, 75, 76
  monozygotic, 75, 76
  decline in DZ rate, 85
  secular changes in rates, 81, 82
uterus, 2, 6, 7
vagina, 2, 7, 121
X-spermatozoa, 87, 113
Y-spermatozoa, 87, 113